950

INTRODUCTION TO THE
PHYSICS OF THE EARTH'S INTERIOR

CAMBRIDGE TOPICS IN MINERAL PHYSICS AND CHEMISTRY

Editors
Dr. Andrew Putnis
Dr. Robert C. Liebermann

Introduction to the physics of the Earth's interior

JEAN-PAUL POIRIER

Institut de Physique de Globe de Paris

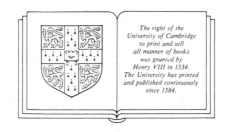

The right of the
University of Cambridge
to print and sell
all manner of books
was granted by
Henry VIII in 1534.
The University has printed
and published continuously
since 1584.

CAMBRIDGE UNIVERSITY PRESS

CAMBRIDGE

NEW YORK PORT CHESTER MELBOURNE SYDNEY

Published by the Press Syndicate of the University of Cambridge
The Pitt Building, Trumpington Street, Cambridge CB2 1RP
40 West 20th Street, New York, NY 10011, USA
10 Stamford Road, Oakleigh, Melbourne 3166, Australia

First published 1991

Printed in Canada

Library of Congress Cataloging-in-Publication Data
Poirier, Jean Paul.
Introduction to the physics of the Earth's interior / Jean-Paul Poirier.
p.   cm. – (Cambridge topics in mineral physics and chemistry : 3)
Includes bibliographical references and index.
ISBN 0-521-38097-9 (hardback). – ISBN 0-521-38801-5 (pbk.)
1. Earth – Interior.   2. Geophysics.   I. Title.   II. Series.
QE509.P64   1991
551.1'1 – dc20   90–25400
                                                      CIP

British Library Cataloguing in Publication data applied for

ISBN 0-521-38097-9 hardback
ISBN 0-521-38801-5 paperback

Contents

Preface

Not so long ago, Geophysics was a part of Meteorology and there was no such thing as Physics of the Earth's interior. Then came Seismology and, with it, the realization that the elastic waves excited by earthquakes, refracted and reflected within the Earth, could be used to probe its depths and gather information on the elastic structure and eventually the physics and chemistry of inaccessible regions down to the center of the Earth.

The basic ingredients are the travel times of various phases, on seismograms recorded at stations all over the globe. Inversion of a considerable amount of data yields a seismological Earth model, that is, essentially a set of values of the longitudinal and transverse elastic wave velocities for all depths. It is well known that the velocities depend on the elastic moduli and the density of the medium in which the waves propagate; the elastic moduli and the density, in turn, depend on the crystal structure and chemical composition of the constitutive minerals and on pressure and temperature. To extract from velocity profiles self-consistent information on the Earth's interior such as pressure, temperature, and composition as a function of depth, one needs to know, or at least estimate, the values of the physical parameters of the high-pressure and high-temperature phases of the candidate minerals, and relate them, in the framework of thermodynamics, to the Earth's parameters.

Physics of the Earth's interior has expanded from there to become a recognized discipline within solid earth geophysics, and an important part of the current geophysical literature can be found under such key words as "equation of state," "Grüneisen parameter," "adiabaticity," "melting curve," "electrical conductivity," and so on.

The problem, however, is that, although most geophysics textbooks devote a few paragraphs, or even a few chapters, to the basic concepts of the physics of solids and its applications, there still is no self-contained book that offers the background information needed by the graduate student or the nonspecialist geophysicist to understand an increasing portion of the literature as well as to assess the weight of physical arguments from various parties in current controversies about the structure, composition, or temperature of the deep Earth.

The present book has the, admittedly unreasonable, ambition to fulfill this role. Starting as a primer, and giving at length all the important demonstrations, it should lead the reader, step by step, to the most recent developments in the literature. The book is primarily intended for graduate or senior undergraduate students in physical earth sciences but it is hoped that it can also be useful to geophysicists interested in getting acquainted with the mineral physics foundations of the phenomena they study.

In the first part, the necessary background in thermodynamics of solids is succinctly given in the framework of linear relations between intensive and extensive quantities. Elementary solid-state theory of vibrations in solids serves as a basis to introduce Debye's theory of specific heat and anharmonicity. Many definitions of Grüneisen's parameter are given and compared.

The background is used to explain the origin of the various equations of state (Murnaghan, Birch–Murnaghan, etc.). Velocity–density systematics and Birch's law lead to seismic equations of state. Shock-wave equations of state are also briefly considered. Tables of recent values of thermodynamic and elastic parameters of the most important mantle minerals are given. The effect of pressure on melting is introduced in the framework of anharmonicity, and various melting laws (Lindemann, Kraut–Kennedy, etc.) are given and discussed. Transport properties of materials – diffusion and viscosity of solids and of liquid metals, electrical and thermal conductivity of solids – are important in understanding the workings of the Earth; a chapter is devoted to them.

The last chapter deals with the application of the previous ones to the determination of seismological, thermal, and compositional Earth models.

An abundant bibliography, including the original papers and the most recent contributions, experimental or theoretical, should help the reader to go further than the limited scope of the book.

It is a pleasure to thank all those who helped make this book come into being: first of all, Bob Liebermann, who persuaded me to write it and suggested improvements in the manuscript; Joël Dyon, who did a splendid job on the artwork; Claude Allègre, Vincent Courtillot, François Guyot, Jean-Louis Mouël, and Jean-Paul Montagner, who read all or parts of the manuscript and provided invaluable comments and suggestions; and last but not least, Carol, for everything.

Introduction

The interior of the Earth is a problem at once fascinating and baffling, as one may easily judge from the vast literature and the few established facts concerning it.

F. Birch, J. Geophys. Res., 57, 227 (1952)

This book is about the inaccessible interior of the Earth. Indeed, it is because it is inaccessible, hence known only indirectly and with a low resolving power, that we can talk of the physics of the interior of the Earth. The Earth's crust has been investigated for many years by geologists and geophysicists of various persuasions; as a result, it is known with such a wealth of detail that it is almost meaningless to speak of the crust as if it were a homogeneous medium endowed with averaged physical properties, in a state defined by simple temperature and pressure distributions. We have the physics of earthquake sources, of sedimentation, of metamorphism, of magnetic minerals, and so forth but no physics of the crust.

Below the crust, however, begins the realm of inner Earth, less well known and apparently simpler: a world of successive homogeneous spherical shells, with a radially symmetrical distribution of density and under a predominantly hydrostatic pressure. To these vast regions, we can apply macroscopic phenomenologies such as thermodynamics or continuum mechanics, deal with energy transfers using the tools of physics, and obtain Earth models, seismological, thermal, or compositional. These models, such as they were until, say, about 1950, accounted for the gross features of the interior of the Earth: a silicate mantle whose density increased with depth as it was compressed, with a couple of seismological discontinuities inside, a liquid iron core where convection currents generated the Earth's magnetic field, and a small solid inner core.

The physics of the interior of the Earth arguably came of age in the 1950s, when, following Bridgman's tracks, Birch at Harvard University and Ringwood at the Australian National University started investigating the high-pressure properties and transformations of the silicate minerals. Large-volume multi-anvil presses were developed in Japan (see Akimoto 1987) and diamond-anvil cells were developed in the United States (see Bassett 1977), allowing the synthesis of minerals at the static pressures of the

1

lower mantle, while shock-wave techniques (see Ahrens 1980) produced high dynamic pressures. It turns out, fortunately, that the wealth of mineral architecture that we see in the crust and uppermost mantle reduces to a few close-packed structures at very high pressures.

It is now possible to use the arsenal of modern methods (e.g., spectroscopies from the infrared to the hard X rays generated in synchrotrons) to investigate the physical properties of the materials of the Earth at very high pressures, thus giving a firm basis to the averaged physical properties of the inner regions of the Earth deduced from seismological or geomagnetic observations and allowing the setting of constraints on the energetics of the Earth.

It is the purpose of this book to introduce the groundwork of condensed matter physics, which has allowed, and still allows, the improvement of Earth models. Starting with the indispensable, if somewhat arid, phenomenological background of thermodynamics of solids and continuum mechanics, we will relate the macroscopic observables to crystalline physics; we will then deal with melting, phase transitions, and transport properties before trying to synthetically present the Earth models of today.

The role of laboratory experimentation cannot be overestimated. It is, however, beyond the scope of this book to present the experimental techniques, but references to review articles will be given.

In a book such as this one, which topic to include or reject is largely a matter of personal, hence debatable, choice. I give only a brief account of the phase transitions of minerals in a paragraph that some readers may well find somewhat skimpy; I choose to do so because this active field is in rapid expansion and I prefer outlining the important results and giving recent references to running the risk of confusing the reader. Also, little is known yet about the mineral reactions in the transition zone and the lower mantle, so I deal only with the polymorphic, isochemical transitions of the main mantle minerals, thus keeping well clear of the huge field of experimental petrology.

It is hoped that this book may help with the understanding of how condensed matter physics may be of use in improving Earth models. It will also probably become clear that the simplicity of the inner Earth is only apparent; with the progress of laboratory experimental techniques as well as observational seismology, geochemistry, and geomagnetism, we may perhaps expect that someday "physics of the interior of the Earth" will make as little sense as "physics of the crust."

1

Background of thermodynamics
of solids

1.1 Extensive and intensive conjugate quantities

The physical quantities used to define the state of a system can be scalar
(e.g., volume, hydrostatic pressure, number of moles of constituent), vectorial (e.g., electric or magnetic field), or tensorial (e.g., stress or strain).
In all cases, one may distinguish extensive and intensive quantities. The
distinction is most obvious for scalar quantities: extensive quantities are
size-dependent (e.g., volume, entropy) and intensive quantities are not
(e.g., pressure, temperature).

Conjugate quantities are such that their product (scalar or contracted
product for vectorial and tensorial quantities) has the dimension of energy
or energy per unit volume, depending on the definition of the extensive
quantities (Table 1.1). By analogy with the expression of mechanical work
as the product of a force by a displacement, the intensive quantities are
also called *generalized forces* and the extensive quantities, *generalized
displacements*.

If the state of a single-phased system is defined by N extensive quantities e_k and N intensive quantities i_k, the differential increase in energy
per unit volume of the system for a variation of e_k is

$$dU = \sum_k i_k \, de_k \tag{1.1}$$

The intensive quantities can therefore be defined as partial derivatives
of the energy with respect to their conjugate quantities,

$$i_k = \frac{\partial U}{\partial e_k} \tag{1.2}$$

For the extensive quantities, we have to introduce the Gibbs potential

$$G = U - \sum i_k e_k \tag{1.3}$$

$$dG = \sum i_k \, de_k - d \sum i_k e_k = - \sum e_k \, di_k \tag{1.4}$$

3

Table 1.1. *Some examples of conjugate quantities*

Intensive quantities	i_k	Extensive quantities	e_k
Temperature	T	Entropy	S
Pressure	P	Volume	V
Chemical potential	μ	Number of moles	n
Electric field	$\mathbf{E}$	Displacement	$\mathbf{D}$
Magnetic field	$\mathbf{H}$	Induction	$\mathbf{B}$
Stress	σ	Strain	ϵ

and we have

$$e_k = -\frac{\partial G}{\partial i_k} \tag{1.5}$$

Conjugate quantities are linked by *constitutive relations* that express the response of the system in terms of one quantity, when its conjugate is made to vary. The relations are usually taken to be linear and the proportionality coefficient is a *material constant* (e.g., elastic moduli in Hooke's law).

In general, starting from a given state of the system, if all the intensive quantities are arbitrarily varied, the extensive quantities will vary (and vice versa). As a first approximation, the variations are taken to be linear and systems of linear equations are written (Zwikker 1954):

$$de_k = \kappa_{k1} di_1 + \kappa_{k2} di_2 + \cdots + \kappa_{kn} di_n \tag{1.6}$$

or

$$di_k = K_{k1} de_1 + K_{k2} de_2 + \cdots + K_{kn} de_n \tag{1.7}$$

The constants

$$\kappa_{kl} = \left(\frac{\partial e_k}{\partial i_l}\right)_{i1,\ldots,in,\,\text{except } il} \tag{1.8}$$

are called *compliances* (e.g., compressibility), and the constants

$$K_{lk} = \left(\frac{\partial i_l}{\partial e_k}\right)_{e1,\ldots,en,\,\text{except } ek} \tag{1.9}$$

are called *stiffnesses* (e.g., bulk modulus).

Note that, in general,

$$K_{lk} \neq \frac{1}{\kappa_{kl}}$$

The linear approximation, however, holds only locally for small values of the variations about the reference state, and we will see that, in many instances, it cannot be used. It is particularly true for the relation between pressure and volume, deep inside the Earth: very high pressures create finite strains and the linear relation (Hooke's law) is not valid over such a wide range of pressure. One, then, has to use more sophisticated equations of state.

1.2 Thermodynamic potentials

The energy of a thermodynamic system is a state function, that is, its variation depends only on the initial and final states and not on the path from the one to the other. The energy can be expressed as various potentials according to which extensive or intensive quantities are chosen as independent variables. The most commonly used are the *internal energy* E, for the variables volume and entropy, the *enthalpy* H, for pressure and entropy, the *Helmholtz free energy* F, for volume and temperature, and the *Gibbs free energy* G, for pressure and temperature:

$$E \tag{1.10}$$

$$H = E + PV \tag{1.11}$$

$$F = E - TS \tag{1.12}$$

$$G = H - TS \tag{1.13}$$

The differentials of these potentials are total exact differentials

$$dE = T\,dS - P\,dV \tag{1.14}$$

$$dH = T\,dS + V\,dP \tag{1.15}$$

$$dF = -S\,dT - P\,dV \tag{1.16}$$

$$dG = -S\,dT + V\,dP \tag{1.17}$$

The extensive and intensive quantities can therefore be expressed as partial derivatives according to (1.2) and (1.5):

$$T = \left(\frac{\partial E}{\partial S}\right)_V = \left(\frac{\partial H}{\partial S}\right)_P \tag{1.18}$$

$$S = -\left(\frac{\partial F}{\partial T}\right)_V = -\left(\frac{\partial G}{\partial T}\right)_P \tag{1.19}$$

$$P = -\left(\frac{\partial E}{\partial V}\right)_S = -\left(\frac{\partial F}{\partial V}\right)_T \tag{1.20}$$

$$V = \left(\frac{\partial H}{\partial P}\right)_S = \left(\frac{\partial G}{\partial P}\right)_T \tag{1.21}$$

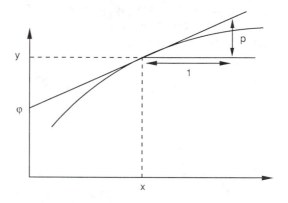

Figure 1.1. Legendre's transformation: the curve $y = f(x)$ is defined as the envelope of its tangents of equation $\phi = y - px$.

In accordance with the usual convention, a subscript is used to identify the independent variable that stays fixed.

From the first principle of thermodynamics, the differential of internal energy dE of a closed system is the sum of a heat term $dQ = T dS$ and a mechanical work term $dW = -P dV$. The internal energy is therefore the most physically understandable thermodynamic potential; unfortunately, its differential is expressed in terms of the independent variables entropy and volume, which are not the most convenient in many cases. The existence of the other potentials H, F, and G has no justification other than their being more convenient in specific cases. Their expression is not gratuitous, nor does it have some deep and hidden meaning. It is just the result of a mathematical transformation (Legendre transformation), whereby a function of one or more variables can be expressed in terms of its partial derivatives, which become independent variables (see Callen 1985).

The idea can be easily understood, using as an example a function y of a variable x: $y = f(x)$. The function is represented by a curve in the (x, y) plane (Fig. 1.1), and the slope of the tangent to the curve at point (x, y) is: $p = dy/dx$. The tangent cuts the y-axis at the point $(0, \varphi)$ and its equation is: $\varphi = y - px$. This equation represents the curve defined as the envelope of its tangents, that is, as a function of the derivative p of $y(x)$.

In our case, we deal with a surface that can be represented as the envelope of its tangent planes. Supposing we want to express $E(S, V)$ in terms of T and P. We write the equation of the tangent plane

$$\varphi = E - \left(\frac{\partial E}{\partial V}\right)_S V - \left(\frac{\partial E}{\partial S}\right)_V S = E + PV - TS = G$$

In geophysics, we are mostly interested in the variables T and P. We will therefore mostly use the Gibbs free energy.

1.3 Maxwell's relations. Stiffnesses and compliances

The potentials are functions of state and their differentials are total exact differentials. The second derivatives of the potentials with respect to the independent variables do not depend on the order in which the successive derivatives are taken. Starting from equations (1.18)–(1.21), we therefore obtain *Maxwell's relations:*

$$-\left(\frac{\partial S}{\partial P}\right)_T = \left(\frac{\partial V}{\partial T}\right)_P \tag{1.22}$$

$$\left(\frac{\partial S}{\partial V}\right)_T = \left(\frac{\partial P}{\partial T}\right)_V \tag{1.23}$$

$$\left(\frac{\partial T}{\partial P}\right)_S = \left(\frac{\partial V}{\partial S}\right)_P \tag{1.24}$$

$$\left(\frac{\partial T}{\partial V}\right)_S = -\left(\frac{\partial P}{\partial S}\right)_V \tag{1.25}$$

Other relationships between the second partial derivatives can be obtained, using the chain rule for the partial derivatives of a function $f(x, y, z) = 0$:

$$\left(\frac{\partial x}{\partial y}\right)_z \left(\frac{\partial y}{\partial z}\right)_x \left(\frac{\partial z}{\partial x}\right)_y = -1 \tag{1.26}$$

For instance, assuming a relation $f(P, V, T) = 0$, we have

$$\left(\frac{\partial V}{\partial T}\right)_P = -\left(\frac{\partial V}{\partial P}\right)_T \left(\frac{\partial P}{\partial T}\right)_V \tag{1.27}$$

With Maxwell's relations, the chain rule yields relations among all derivatives of the intensive and extensive variables with respect to one another (Table 1.2).

We must be aware that Maxwell's relations involved only conjugate quantities, but that by using the chain rule, we introduce derivatives of intensive or extensive quantities with respect to nonconjugate quantities. These will have a meaning only if we consider cross-couplings between fields (e.g., thermoelastic coupling, see Sec. 2.3) and the material constants correspond to second-order effects (e.g., thermal expansion).

In Zwikker's notation, the second derivatives of the potentials are stiffnesses and compliances (Sec. 1.1)

$$K_{lk} = \frac{\partial i_l}{\partial e_k} = \frac{\partial^2 U}{\partial e_l \, \partial e_k} \tag{1.28}$$

$$\kappa_{kl} = \frac{\partial e_k}{\partial i_l} = -\frac{\partial^2 G}{\partial i_k \, \partial i_l} \tag{1.29}$$

Table 1.2. *Derivatives of extensive* (S, V) *and intensive* (T, P) *quantities*

$$\left(\frac{\partial S}{\partial T}\right)_V = \frac{C_V}{T} \qquad \left(\frac{\partial S}{\partial V}\right)_T = \alpha K_T \qquad \left(\frac{\partial S}{\partial P}\right)_V = \frac{C_P}{\alpha K_S T}$$

$$\left(\frac{\partial S}{\partial T}\right)_P = \frac{C_P}{T} \qquad \left(\frac{\partial S}{\partial V}\right)_P = \frac{C_P}{\alpha V T} \qquad \left(\frac{\partial S}{\partial P}\right)_T = -\alpha V$$

$$\left(\frac{\partial T}{\partial S}\right)_V = \frac{T}{C_V} \qquad \left(\frac{\partial T}{\partial V}\right)_S = -\frac{\alpha K_S T}{C_P} \qquad \left(\frac{\partial T}{\partial P}\right)_V = \frac{1}{\beta P}$$

$$\left(\frac{\partial T}{\partial S}\right)_P = \frac{T}{C_P} \qquad \left(\frac{\partial T}{\partial V}\right)_P = \frac{1}{\alpha V} \qquad \left(\frac{\partial T}{\partial P}\right)_S = \frac{\alpha V T}{C_P}$$

$$\left(\frac{\partial P}{\partial T}\right)_V = \alpha K_T \qquad \left(\frac{\partial P}{\partial V}\right)_S = -\frac{K_S}{V} \qquad \left(\frac{\partial P}{\partial S}\right)_T = -\frac{1}{\alpha V}$$

$$\left(\frac{\partial P}{\partial T}\right)_S = \frac{C_P}{\alpha V T} \qquad \left(\frac{\partial P}{\partial V}\right)_T = -\frac{K_T}{V} \qquad \left(\frac{\partial P}{\partial S}\right)_V = \frac{\alpha K_S T}{C_P}$$

$$\left(\frac{\partial V}{\partial T}\right)_S = -\frac{C_P}{\alpha K_S T} \qquad \left(\frac{\partial V}{\partial P}\right)_S = -\frac{V}{K_S} \qquad \left(\frac{\partial V}{\partial S}\right)_T = \frac{1}{\alpha K_T}$$

$$\left(\frac{\partial V}{\partial T}\right)_P = \alpha V \qquad \left(\frac{\partial V}{\partial P}\right)_T = -\frac{V}{K_T} \qquad \left(\frac{\partial V}{\partial S}\right)_P = \frac{\alpha V T}{C_P}$$

It follows, since the order of differentiation can be reversed, that

$$K_{lk} = K_{kl} \tag{1.30}$$

$$\kappa_{kl} = \kappa_{lk} \tag{1.31}$$

Inspection of Table 1.2 shows that, depending on which variables are kept constant when the derivative is taken, we define isothermal, K_T, and adiabatic, K_S, bulk moduli and isobaric, C_P, and isochoric, C_V, specific heats. We must note here that the adiabatic bulk modulus is a stiffness, whereas the isothermal bulk modulus is the reciprocal of a compliance, hence they are not equal (Sec. 1.1); similarly, the isobaric specific heat is a compliance, whereas the isochoric specific heat is the reciprocal of a stiffness.

Table 1.2 contains extremely useful relations, involving the thermal and mechanical material constants, which we will use throughout this book. Note that, here and throughout the book, V is the specific volume. We will also use the specific mass ρ, with $V\rho = 1$. Often loosely called density, the specific mass is numerically equal to density only in unit systems in which the specific mass of water is equal to unity.

2

Elastic moduli

2.1 Background of linear elasticity

We will rapidly review here the most important results and formulas of linear (Hookean) elasticity. For a complete treatment of elasticity, the reader is referred to the classic books on the subject (Love 1944; Brillouin 1960; Nye 1957). See also Means (1976) for a clear treatment of stress and strain at the beginner's level.

Let us start with the definition of infinitesimal strain (a general definition of finite strain will be given in Chapter 4). We define the tensor of infinitesimal strain ϵ_{ij}, $i, j = 1, 2, 3$, as the symmetrical part of the displacement gradient tensor $\partial u_i / \partial x_j$, where the u_is are the components of the displacement vector of a point of coordinates x_j (Fig. 2.1),

$$\epsilon_{ij} = \frac{1}{2} \left(\frac{\partial u_i}{\partial x_j} + \frac{\partial u_j}{\partial x_i} \right) \tag{2.1}$$

The trace of the strain tensor is the *dilatation* (positive or negative)

$$\mathrm{Tr}\, \epsilon_{ij} = \sum_k \epsilon_{kk} = \frac{\partial u_1}{\partial x_1} + \frac{\partial u_2}{\partial x_2} + \frac{\partial u_3}{\partial x_3} = \mathrm{div}\, \mathbf{u} \cong \frac{\Delta V}{V} \tag{2.2}$$

The components σ_{ij} of the stress tensor are defined in the following way: Let us consider a volume element around a point in a solid submitted to surface and/or body forces. If we cut the volume element by a plane normal to the coordinate axis i and remove the part of the solid on the side of the positive axis, its action on the volume element can be replaced by a force, whose components along the j axis are σ_{ij} (Fig. 2.2). In the absence of body torque, the stress tensor is symmetrical.

The trace of the stress tensor is equal to three times the *hydrostatic pressure:*

$$\mathrm{Tr}\, \sigma_{ij} = \sigma_{11} + \sigma_{22} + \sigma_{33} = 3P \tag{2.3}$$

9

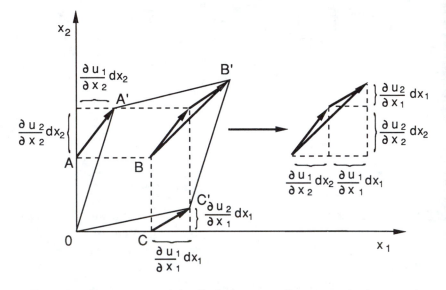

Figure 2.1. Components of the displacement gradient tensor in the case of infinitesimal plane strain. The components of the strain tensor are

$$\epsilon_{11} = \partial u_1 / \partial x_1, \qquad \epsilon_{22} = \partial u_2 / \partial x_2, \qquad \epsilon_{12} = \tfrac{1}{2}(\partial u_1 / \partial x_{12} + \partial u_2 / \partial x_1) = \epsilon_{21}$$

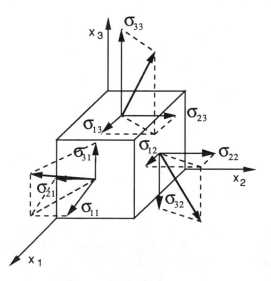

Figure 2.2. Components of the stress tensor σ_{ij}. The bold vectors represent the force per unit area exerted on the volume element by the (removed) part of the solid on the positive side of the normal to the corresponding plane.

Hence the hydrostatic pressure is

$$P = \frac{1}{3} \sum_k \sigma_{kk} \tag{2.4}$$

2.2 Elastic constants and moduli

For an isotropic, homogeneous solid and infinitesimal strains, there is a linear constitutive relation between the second-order tensors of stress and strain, that expresses the response of an elastic solid to the application of stress or strain, starting from an initial, "natural," stress and strain–free state.

▶
$$\sigma_{ij} = \sum_{kl} c_{ijkl} \epsilon_{kl} \tag{2.5}$$

This is Hooke's law. The fourth-order symmetrical tensor c_{ijkl} is the *elastic constants* tensor. Due to the fact that the stress and strain tensors are symmetrical, the most general elastic constants tensor has only twenty-one nonzero independent components. For crystals, the number of independent elastic constants decreases as the symmetry of the crystalline system increases, and it reduces to three for the cubic system

$$c_{1111} = c_{11}, \qquad c_{1122} = c_{12}, \qquad c_{2323} = c_{44}$$

The elastic constants are usually expressed in contracted notation, pairs of indices being replaced by one index according to the correspondance rule:

$$11 \to 1, \quad 22 \to 2, \quad 33 \to 3, \quad 23 = 32 \to 4, \quad 13 = 31 \to 5, \quad 12 = 21 \to 6$$

In what follows, we will mostly give examples relative to cubic crystals, for the sake of simplicity and also because many of the most important minerals of the deep Earth are cubic (spinel, garnet, magnesiowüstite, ideal silicate perovskites).

For an isotropic system (e.g., an aggregate of crystals in various random orientations), the number of independent elastic constants reduces to two. Hooke's law is then conveniently expressed as

▶
$$\sigma_{ij} = \lambda \delta_{ij} \sum_k \epsilon_{kk} + 2\mu \epsilon_{ij} \tag{2.6}$$

where δ_{ij} is equal to 1 if $i = j$ and to zero if $i \neq j$, $\sum_k \epsilon_{kk} = \Delta V/V$ is the trace of the strain tensor, and λ and μ are the two independent *Lamé constants,* defined by

$$\mu = c_{44} \quad \text{(the } shear\ modulus\text{)}$$

and

$$\lambda + 2\mu = c_{11}$$

Hence

$$\lambda = c_{12} = c_{11} - 2c_{44}$$

Note that c_{11}, c_{12}, and c_{44} here are the three nonindependent elastic constants of the isotropic aggregate, not the three independent constants of cubic crystals.

The elastic properties of an isotropic material can be described by *elastic moduli,* which consist of any two convenient functions of λ and μ.

The elastic moduli most currently used in solid earth geophysics are (see Weidner (1987) for a recent review of the experimental methods of determination of the elastic moduli):

The *shear modulus* μ. For cubic crystals, there are two shear moduli: c_{44} and $c' = \frac{1}{2}(c_{11} - c_{12})$, corresponding to the shear along the {100} (cube) and {110} (diagonal) planes, respectively.

The *bulk modulus* or *incompressibility* K, defined (Table 1.2) by

$$K = -V\frac{dP}{dV} = -\frac{dP}{d \ln V} \tag{2.7}$$

In linear elasticity, when a pressure P is applied to a solid in the natural state, the corresponding relative volume change is given by

$$\frac{\Delta V}{V} = -\frac{P}{K} \tag{2.8}$$

Hence, from (2.6),

$$K = \frac{3\lambda + 2\mu}{3} \tag{2.9}$$

Poisson's ratio v, defined in a regime of uniaxial stress σ_{11} as minus the ratio of the strain normal to the stress axis, $\epsilon_{22} = \epsilon_{33}$, to the strain along the stress axis, ϵ_{11} (i.e., ratio of thinning to elongation or thickening to contraction):

$$v = \frac{-\epsilon_{22}}{\epsilon_{11}} = \frac{-\epsilon_{33}}{\epsilon_{11}} \tag{2.10}$$

Poisson's ratio, being dimensionless, is not strictly speaking a modulus, but it is a combination of elastic moduli and it can be used, together with any one modulus to completely define the elastic properties of a body. Indeed, using (2.6) and writing $\sigma_{22} = \sigma_{33} = 0$, we obtain

$$v = \frac{\lambda}{2(\lambda + \mu)} \tag{2.11}$$

and, with (2.9),

$$\nu = \frac{3K-2\mu}{2(3K+\mu)} = \frac{3(K/\mu)-2}{2(3K/\mu+1)} \qquad (2.12)$$

In many cases, especially in the Earth's crust, it so happens that $\lambda = \mu$, that is, there is only one independent elastic modulus (Cauchy solid); then $\nu = 0.25$.

If the solid is incompressible ($K = \infty$), then, from (2.12), $\nu = 0.5$. The same result can of course be obtained with the definition of ν, by writing $\Delta V/V = \epsilon_1 + 2\epsilon_2 = 0$. Note that for a liquid $\mu = 0$, hence we also have $\nu = 0.5$, but that does not mean that the liquid is incompressible.

For an infinitely compressible solid ($K = 0$), we would have $\nu = -1$.

We therefore have the bounds on Poisson's ratio

$$-1 < \nu < 0.5 \qquad (2.13)$$

Poisson's ratio is especially interesting in geophysics, because it can be expressed as a function of the ratio v_P/v_S of the velocities of the longitudinal (P) and transverse (S) elastic waves only. We have

$$v_P = \left(\frac{\lambda+2\mu}{\rho}\right)^{1/2} = \left(\frac{K+4\mu/3}{\rho}\right)^{1/2} \qquad (2.14)$$

$$v_S = \left(\frac{\mu}{\rho}\right)^{1/2} \qquad (2.15)$$

Hence

$$\frac{v_P}{v_S} = \left(\frac{\lambda}{\mu+2}\right)^{1/2} \qquad (2.16)$$

From (2.16) and (2.11), we have

$$\nu = \frac{(v_P/v_S)^2-2}{2[(v_P/v_S)^2-1]} \qquad (2.17)$$

The condition $\nu = 0.25$ corresponds to $v_P = v_S\sqrt{3}$, which is frequently obtained in the crust.

Let us remind the reader here that equations (2.14) and (2.15) can be derived from Newton's equation of motion of a unit volume element of a continuum medium:

$$\rho\frac{\partial^2 \mathbf{u}}{\partial t^2} = \mathbf{F} \qquad (2.18)$$

where $\mathbf{u}$ is the displacement vector, ρ the specific mass, and $\mathbf{F}$ the force that balances the stress on the volume element, given by

$$F_i = \sum_j \frac{\partial \sigma_{ij}}{\partial x_j} \qquad (2.19)$$

We will here write the equation of motion in the simple case of a longitudinal wave, propagating in the x_1 direction ($u_1 = u$, $u_2 = u_3 = 0$) and a shear wave polarized along x_2 and propagating along x_1 ($u_1 = u$, $u_2 = u_3 = 0$, $\partial u_1 / \partial x_2 = \partial u / \partial x$, $\partial u_1 / \partial x_1 = \partial u_1 / \partial x_3 = 0$).

From (2.1), (2.6), (2.18), and (2.19), we have for the longitudinal wave

$$\sigma_{11} = (\lambda + 2\mu) \frac{\partial u_1}{\partial x_1}$$

and

$$\rho \frac{\partial^2 u}{\partial t^2} = (\lambda + 2\mu) \frac{\partial^2 u}{\partial x^2} \tag{2.20}$$

and for the shear wave

$$\sigma_{12} = \mu \frac{\partial u_1}{\partial x_2}$$

and

$$\rho \frac{\partial^2 u}{\partial t^2} = \mu \frac{\partial^2 u}{\partial x^2} \tag{2.21}$$

The wave equations (2.20) and (2.21) correspond to waves propagating with velocities given by (2.14) and (2.15), respectively.

Here is a good opportunity to introduce the *seismic parameter* Φ, which we will frequently use later on,

$$\blacktriangleright \qquad \Phi = \frac{K}{\rho} \tag{2.22}$$

It is related to v_Φ, the propagation velocity of the hydrostatic part of the strain (dilatation), often called bulk velocity or hydrodynamic velocity, and given by

$$v_\Phi = \left(\frac{K}{\rho} \right)^{1/2} = \left(\frac{3\lambda + 2\mu}{3\rho} \right)^{1/2} \tag{2.23}$$

Hence

$$\blacktriangleright \qquad \Phi = v_P^2 - \tfrac{4}{3} v_S^2 \tag{2.24}$$

Note that

$$v_\Phi < v_P \quad \text{in solids, because} \quad \lambda + 2\mu/3 < \lambda + 2\mu \quad \text{for } \mu > 0$$

$$v_\Phi = v_P \quad \text{in liquids, because} \quad \mu = 0 \text{ (the strain is purely dilatational)}$$

We can find another useful expression for Φ from the definition of K (2.7):

$$K = -\frac{dP}{d \ln V} = \frac{dP}{d \ln \rho} = \rho \frac{dP}{d\rho} \tag{2.25}$$

where $\rho = 1/V$ is the specific mass.

Hence

$$\Phi = \frac{dP}{d\rho} \tag{2.26}$$

The bulk modulus K is, by definition, isotropic. The average bulk modulus of a single-phased aggregate of anisotropic crystals is therefore the same as the bulk modulus of the individual crystals, and it can easily be found from the experimentally determined elastic constants.

For cubic crystals:

$$K = \frac{c_{11} + 2c_{12}}{3} \tag{2.27}$$

The problem of calculating the effective shear moduli of an aggregate from the single crystal elastic constants is, however, much more difficult and, indeed, it has no exact solution; all we know is that the aggregate value lies between two bounds (see Watt, Davies, and O'Connell 1976): a lower bound calculated assuming that the stress is uniform in the aggregate and that the strain is the total sum of all the strains of the individual grains in series (*Reuss bound*), and an upper bound calculated assuming that the strain is uniform and that the stress is supported by the individual grains in parallel (*Voigt bound*). The arithmetic average of the two bounds is often used (*Voigt–Reuss–Hill average*).

Variational methods allow the calculation of the tighter *Hashin–Shtrikman bounds* (Watt et al. 1976; Watt 1988).

For cubic crystals, with elastic constants c_{11}, c_{12}, c_{44}, there are two shear moduli c and c' corresponding to shear on the {100} and {110} planes, respectively,

$$c = c_{44}$$

$$c' = \tfrac{1}{2}(c_{11} - c_{12})$$

The effective Reuss and Voigt shear moduli of a single-phased aggregate are

$$\mu_R = \frac{15}{6/c' + 9/c} \tag{2.28a}$$

$$\mu_V = \frac{1}{5}(2c' + 3c) \tag{2.28b}$$

Expressions for the effective moduli of aggregates of crystals with lower symmetry can be found in Sumino and Anderson (1984).

The lower and upper Hashin–Shtrikman bounds for an aggregate of cubic crystals are

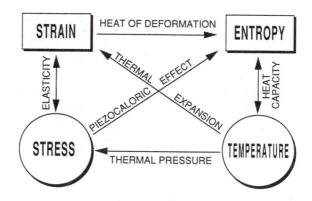

Figure 2.3. Schema of the coupling between thermal and mechanical variables (after Nye 1957).

$$\mu_L = c' + \frac{3}{5}(c - c' - 4\beta) \tag{2.29a}$$

$$\mu_U = c + \frac{2}{5}(c' - c - 6\beta') \tag{2.29b}$$

with

$$\beta = \frac{3}{5} \frac{K + 2c'}{c'(3K + 4c')}$$

$$\beta' = \frac{3}{5} \frac{K + 2c}{c(3K + 4c)}$$

2.3 Thermoelastic coupling

2.3.1 Generalities

All the extensive and intensive variables, conjugate or not, can be cross-coupled in many ways and the couplings are responsible for a variety of first- and second-order physical effects, for example, thermoelastic or piezoelectric effects (see Nye 1957). We will deal here only with thermoelastic coupling (Fig. 2.3) and derive the expressions for the isothermal and adiabatic bulk moduli.

2.3.2 Isothermal and adiabatic moduli

Let us assume that the intensive variables σ_{ij} and T depend only on the two extensive variables ϵ_{kl} and S and that we can write the coupled equations for the differentials (Nye 1957):

$$d\sigma_{ij} = \left(\frac{\partial \sigma_{ij}}{\partial \epsilon_{kl}}\right)_S d\epsilon_{kl} + \left(\frac{\partial \sigma_{ij}}{\partial S}\right)_\epsilon dS \qquad (2.30)$$

$$dT = \left(\frac{\partial T}{\partial \epsilon_{kl}}\right)_S d\epsilon_{kl} + \left(\frac{\partial T}{\partial S}\right)_\epsilon dS \qquad (2.31)$$

Let us consider only the simple scalar case (geophysically relevant) of hydrostatic pressure $\sigma_{ij} = \delta_{ij}P$ and isotropic compression $\epsilon = \Delta V/V$.

We can then write (2.30) and (2.31) as

$$dP = \left(\frac{\partial P}{\partial \epsilon}\right)_S d\epsilon + \left(\frac{\partial P}{\partial S}\right)_\epsilon dS \qquad (2.32)$$

$$dT = \left(\frac{\partial T}{\partial \epsilon}\right)_S d\epsilon + \left(\frac{\partial T}{\partial S}\right)_\epsilon dS \qquad (2.33)$$

Dividing both sides of (2.32) by $d\epsilon$, we get

$$\frac{dP}{d\epsilon} = \left(\frac{\partial P}{\partial \epsilon}\right)_S + \left(\frac{\partial P}{\partial S}\right)_\epsilon \frac{dS}{d\epsilon} \qquad (2.34)$$

Let us consider the isothermal case, and assume $dT = 0$ in (2.33). We find

$$\frac{dS}{d\epsilon} = -\left(\frac{\partial T}{\partial \epsilon}\right)_S \bigg/ \left(\frac{\partial T}{\partial S}\right)_\epsilon \qquad (2.35)$$

Carrying (2.35) into (2.34), we get

$$\left(\frac{dP}{d\epsilon}\right)_T = \left(\frac{\partial P}{\partial \epsilon}\right)_S - \left(\frac{\partial P}{\partial S}\right)_\epsilon \left(\frac{\partial S}{\partial T}\right)_\epsilon \left(\frac{\partial T}{\partial \epsilon}\right)_S$$

$$\left(\frac{dP}{d\epsilon}\right)_T = \left(\frac{\partial P}{\partial \epsilon}\right)_S - \left(\frac{\partial P}{\partial T}\right)_\epsilon \left(\frac{\partial T}{\partial \epsilon}\right)_S \qquad (2.36)$$

Now,

$$\left(\frac{dP}{d\epsilon}\right)_T = K_T \quad \text{[isothermal bulk modulus]}$$

$$\left(\frac{dP}{d\epsilon}\right)_S = K_S \quad \text{[adiabatic bulk modulus]}$$

and

$$\left(\frac{\partial T}{\partial \epsilon}\right)_S = -\left(\frac{\partial T}{\partial V}\right)_S V$$

Hence

$$K_T - K_S = \left(\frac{\partial P}{\partial T}\right)_\epsilon \left(\frac{\partial T}{\partial V}\right)_S V \qquad (2.37)$$

We find in Table 1.2 that

$$\left(\frac{\partial P}{\partial T}\right)_V \equiv \left(\frac{\partial P}{\partial T}\right)_\epsilon = \alpha K_T$$

$$\left(\frac{\partial T}{\partial V}\right)_S = -\frac{\alpha K_S T}{C_P}$$

Hence

$$K_S - K_T = +\alpha K_T T \frac{\alpha K_S V}{C_V} \tag{2.38}$$

The dimensionless parameter in brackets is the *thermodynamic Grüneisen parameter* (see Chapter 3):

$$\gamma_{th} \equiv \frac{\alpha K_T V}{C_V} \tag{2.39}$$

Hence

$$\frac{K_S}{K_T} = 1 + \gamma_{th} \alpha T \tag{2.40}$$

Now, from Table 1.2, we see that

$$\left(\frac{\partial P}{\partial S}\right)_V = \frac{\alpha T K_S}{C_P} \tag{2.41}$$

Using the chain rule (1.26), we find that

$$\left(\frac{\partial T}{\partial S}\right)_V \left(\frac{\partial S}{\partial V}\right)_T = -\left(\frac{\partial T}{\partial V}\right)_S = \frac{\alpha T K_T}{C_V} \tag{2.42}$$

Now, Maxwell's relations give

$$\left(\frac{\partial T}{\partial V}\right)_S = \left(\frac{\partial P}{\partial S}\right)_V \tag{2.43}$$

Hence, from (2.41), (2.42), (2.43),

$$\blacktriangleright \qquad \frac{C_P}{C_V} = \frac{K_S}{K_T} = 1 + \gamma \alpha T \tag{2.44}$$

Incidentally, we note that

$$\blacktriangleright \qquad \gamma_{th} = \frac{\alpha V K_S}{C_P} = \frac{\alpha V K_T}{C_V} \tag{2.45}$$

Zwikker (1954) gives a general formulation for calculating the difference between a stiffness and the reciprocal of a compliance. Starting from the linear equations (1.7) between intensive and extensive quantities

$$di_1 = K_{11} de_1 + K_{12} de_2$$
$$di_2 = K_{21} de_1 + K_{22} de_2 \tag{2.46}$$

and the definition

$$K_{11} = \left(\frac{\partial i_1}{\partial e_1} \right)_{e2} \tag{2.47}$$

To calculate $1/\kappa_{11} = (\partial i_1 / \partial e_1)_{i2}$ we put $di_2 = 0$ in the second equation and we obtain

$$de_2 = - \frac{K_{21}}{K_{22}} de_1 \tag{2.48}$$

Substituting in the first equation (with $K_{12} = K_{21}$) gives

$$di_1 = \left(K_{11} - \frac{K_{12}^2}{K_{22}} \right) de_{11} \tag{2.49}$$

Hence

$$\left(\frac{\partial i_1}{\partial e_1} \right)_{i2} = \frac{1}{\kappa_{11}} = K_{11} - \frac{K_{12}^2}{K_{22}} \tag{2.50}$$

or

$$K_{11} - \frac{1}{\kappa_{11}} = \frac{K_{12}^2}{K_{22}} \tag{2.51}$$

For thermoelastic coupling, if subscript 1 corresponds to the elastic variables and subscript 2 to the thermal variables (i.e., $i_1 = P$, $i_2 = T$, $e_1 = \epsilon$, $e_2 = S$), then

$$K_{11} = K_S, \quad \frac{1}{\kappa_{11}} = K_T, \quad K_{12} = \left(\frac{\partial P}{\partial S} \right)_V, \quad K_{22} = \left(\frac{\partial T}{\partial S} \right)_V$$

For unit volume, (2.51) is equivalent to (2.38) if we take (2.44) into account.

The difference between the bulk modulus at constant temperature and the bulk modulus at constant entropy (adiabatic) is not trivial since the elastic moduli measured in the laboratory by ultrasonic methods are adiabatic, as well as the ones derived from the seismic wave velocities (the transit time of the waves is too short to allow exchange of heat); on the other hand, the elastic moduli relevant to geodynamic processes on the scale of millions of years are evidently isothermal.

It is, however, interesting to remark that the adiabatic and isothermal shear moduli of an isotropic solid are identical to first order. The following hand waving demonstration is borrowed from Brillouin (1940, p. 23).

Let us consider a solid of unit volume, at equilibrium. Its free energy is a minimum, hence, if we impose a dilatation or a compression, the free energy increases

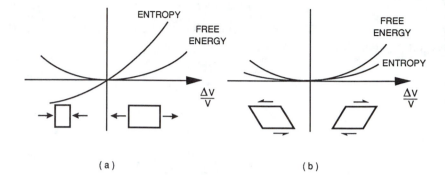

Figure 2.4. Variation of the entropy and free energy in the case of extension-compression (a) and shear (b) elastic deformation (after Brillouin 1940).

in both cases. The free energy curve has a horizontal tangent (Fig. 2.4a). However, due to thermoelastic coupling, the variation of entropy S is not symmetrical: dilatation ($\Delta V/V > 0$) absorbs heat ($\Delta S > 0$), whereas compression ($\Delta V/V < 0$) evolves heat ($\Delta S < 0$). If the entropy is kept constant, the temperature increases on compression and decreases on dilatation. The variation of pressure as a function of $\Delta V/V$ (bulk modulus) is therefore (as seen above) greater for constant entropy than for constant temperature.

Let us now turn to the case of shear strain. For symmetry reasons, at constant temperature, positive and negative shear are equivalent and correspond to an increase in entropy. Free energy and entropy are represented by curves with a minimum and a horizontal tangent (Fig. 2.4b). Hence a shear isothermal transformation is also adiabatic to first order and $\mu_T \cong \mu_S$.

The variation of temperature with reversible adiabatic compression or dilatation is easily found by simple inspection of Table 1.2:

$$\left(\frac{\partial T}{\partial V}\right)_S = -\frac{\alpha K_S T}{C_P} = -\frac{\gamma_{th} T}{V} \tag{2.52}$$

$$\left(\frac{\partial \ln T}{\partial \ln V}\right)_S = -\gamma_{th} \tag{2.53}$$

or

$$\left(\frac{\partial \ln T}{\partial \ln \rho}\right)_S = \gamma_{th} \tag{2.54}$$

We may note that for an adiabatic compression or decompression, we have from (2.53)

$$\left(\frac{T_2}{T_1}\right)_S = \left(\frac{V_1}{V_2}\right)^\gamma \tag{2.55}$$

This relation is known for perfect gases with $\gamma = C_P/C_V - 1$. For solids, we have from (2.44)

$$\gamma = \gamma_{th} = \frac{1}{\alpha T}\left(\frac{C_P}{C_V} - 1\right) \tag{2.56}$$

With the definition $K = dP/d \ln \rho$, we find a useful expression for the variation of temperature with pressure:

$$\left(\frac{\partial T}{\partial P}\right)_S = \gamma_{th}\frac{T}{K_S} \tag{2.57}$$

which, of course, we could have found from Table 1.2 and the definition of γ_{th}.

2.3.3 Thermal pressure

Let us calculate the increase in internal pressure caused by heating a solid at constant volume (*thermal pressure*).

Table 1.2 gives

$$\left(\frac{\partial P}{\partial T}\right)_V = \alpha K_T = \gamma_{th}\frac{C_V}{V} \tag{2.58}$$

Integrating at constant volume and supposing $\gamma_{th} = \text{const.}$, we obtain:

$$P_2 - P_1 \cong \frac{\gamma}{V}\int_{T_1}^{T_2} C_V\, dT = \frac{\gamma}{V}(E_2 - E_1) = \gamma_{th}\frac{\Delta E}{V} \tag{2.59}$$

where E is the internal energy. Hence

$$\blacktriangleright \qquad\qquad \Delta P = \gamma_{th}\frac{\Delta E}{V} \tag{2.60}$$

This is the *Mie-Grüneisen equation of state*, to which we will return later. The Grüneisen parameter is defined here as the coefficient relating the thermal pressure to the thermal energy per unit volume:

$$\gamma_{th} = \frac{V \Delta P}{\Delta E} \tag{2.61}$$

If the volume is not kept constant, we can consider that thermal pressure causes thermal expansion:

$$\left(\frac{\partial P}{\partial T}\right)_V = \alpha K_T$$

is equivalent to

$$\frac{-\Delta P}{K_T} = \frac{\Delta V}{V} = \alpha \Delta T \tag{2.62}$$

This is consistent with the definition of the thermal expansion coefficient

$$\alpha = \left(\frac{\partial \ln V}{\partial T} \right)_P \tag{2.63}$$

We note from (2.61) and (2.62) that the thermal expansion $\Delta V/V$ is proportional to the thermal energy density and from (2.58) that the thermal expansion coefficient is proportional to the specific heat.

Typical values of the thermal expansion coefficient for crystals are:

MgO $\alpha = 3.1 \times 10^{-5} \text{K}^{-1}$

$MgSiO_3$ (enstatite) $\alpha = 2.3 \times 10^{-5} \text{K}^{-1}$

Mg_2SiO_4 (forsterite) $\alpha = 2.5 \times 10^{-5} \text{K}^{-1}$

$MgSiO_3$ (perovskite) 298–381 K $\alpha = 2.2 \times 10^{-5} \text{K}^{-1}$ (Ross and Hazen 1989)

298–840 K $\alpha = 4.0 \times 10^{-5} \text{K}^{-1}$ (Knittle, Jeanloz, and Smith 1987)

Fe $\alpha = 1.5 \times 10^{-5} \text{K}^{-1}$

3

Lattice vibrations

3.1 Generalities

In a crystal at temperatures above absolute zero, atoms vibrate about their equilibrium positions. The crystal can therefore be considered as a collection of oscillators, whose global properties can be calculated. In particular, it will be interesting to determine:

the normal modes of vibration of the crystal;
the dispersion relation, that is, the relation $\omega = f(\mathbf{k})$ between the frequency ω and the wave vector $\mathbf{k}$; and
the vibrational energy.

The vibrational approach is especially fruitful since it allows a synthesis between the thermal and elastic properties and gives a physical basis to thermoelastic coupling. This is due to the fact that the low-frequency, long-wavelength part of the vibrational spectrum corresponds to elastic waves, whereas the high-frequency part corresponds to thermal vibrations. In finite crystals, the lattice vibrations are quantized and behave as quasi-particles: the *phonons*.

In the following section, we will give the elementary basis of the calculations in the simple case of a monatomic lattice. This will be sufficient to introduce the concepts and formulas needed for our purpose. For a more complete and still elementary treatment, the reader is referred to the standard textbooks by Kittel (1967) and Ziman (1965).

3.2 Vibrations of a monatomic lattice

3.2.1 Dispersion curve of an infinite lattice

Let us consider an infinite crystalline lattice formed of only one kind of atoms. Furthermore, let us assume that the lattice is a very simple one

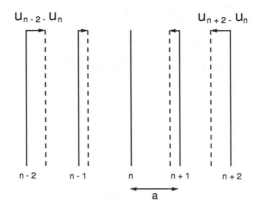

Figure 3.1. Parallel identical lattice planes of an infinite crystalline lattice. The displacement of plane $n+i$ with respect to plane n is $u_{n+i} - u_n$.

and can be described as an infinite stacking of identical, equally spaced, lattice planes. Each atomic plane, of mass M, is labeled by an index n and is connected to all the other planes $n+p$ (p positive or negative can become infinite) by a symmetrical pair-interaction potential $V_{n,n+p}$ (Fig. 3.1).

Let us now consider a longitudinal planar wave propagating normal to the planes (the reasoning would be the same for a shear wave). The displacements $u_{n\pm1}, \dots, u_{n\pm p}$ are counted from an arbitrary origin taken in plane n and they are assumed to be infinitesimal.

Plane n is in the potential of all the other planes $\sum_p V_{n,n+p}$, which can be expanded to second order in powers of $u_n - u_{n+p}$:

$$\sum_p V_{n,n+p} = \sum_p V^0 + \frac{1}{2} \sum_p \frac{\partial^2 V_{n,n+p}}{\partial u_n^2} (u_{n+p} - u_n)^2 + \cdots \qquad (3.1)$$

There is no first-order term since the potential is assumed to be symmetrical; this is an important constraint that will be lifted later on.

The potential well corresponding to a symmetrical potential truncated after the second order is therefore parabolic. This is the *harmonic approximation,* the vibrations of the planes are then harmonic, like those of a mass–spring system, as we will see presently.

Plane n is subjected to a force F_n given by

$$F_n = -\frac{\partial}{\partial u_n} \left(\sum_p V_{n,n+p} \right) = \sum_p \frac{\partial^2 V_{n,n+p}}{\partial u_n^2} (u_{n+p} - u_n) \qquad (3.2)$$

or

$$F_n = \sum_p K_p(u_{n+p} - u_n) \qquad (3.3)$$

with a force constant

$$K_p = \frac{\partial^2 V_{n,n+p}}{\partial u_n^2} = V_p'' \qquad (3.4)$$

The force is linear in displacement as in the case of a harmonic mass-spring system with a force constant K.

If we had only a pair of planes of mass M, Newton's equation would give the equation for a harmonic oscillator

$$M\frac{d^2u}{dt^2} + Ku = 0$$

with a restoring force $F = -dE/du$, which corresponds to a parabolic potential well:

$$E = \frac{Ku^2}{2} + \text{const.}$$

i. Dispersion relations

Let us consider the motion of plane n of mass M in the potential of the other planes. The equation of motion is

$$M\frac{d^2u_n}{dt^2} = \sum_p K_p(u_{n+p} - u_n) \qquad (3.5)$$

Let us look for progressive plane wave solutions:

$$u_n = u^0 \exp i(n\mathbf{k}\cdot\mathbf{a} - \omega t) \qquad (3.6)$$

where $\mathbf{a}$ is the interplanar distance at rest, that is, the period of the lattice (Fig. 3.1).

Let us carry u_n into the equation of motion:

$$-\omega^2 M u^0 \exp i[n\mathbf{k}\cdot\mathbf{a} - \omega t]$$
$$= \sum_p K_p u^0 [\exp i(n+p)\mathbf{k}\cdot\mathbf{a} - \exp in\mathbf{k}\cdot\mathbf{a}] \exp(-i\omega t)$$

or

$$\omega^2 M = -\sum_p K_p \exp(ip\mathbf{k}\cdot\mathbf{a} - 1) \qquad (3.7)$$

All planes being identical, we have $K_p = K_{-p}$ and we can write

$$\omega^2 M = -\sum_{p>0} K_p[\exp(ip\mathbf{k}\cdot\mathbf{a}) + \exp(-ip\mathbf{k}\cdot\mathbf{a}) - 2] \qquad (3.8)$$

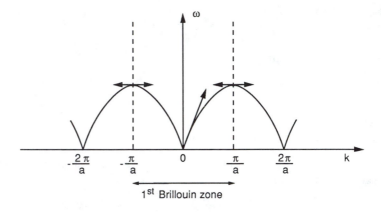

Figure 3.2. Dispersion curve of the lattice of Fig. 3.2. In the center of the Brillouin zone, for long wavelengths, the frequency ω is proportional to the wave number k, hence the group velocity of the lattice waves is equal to the phase velocity of sound. At the edge of the Brillouin zone, the group velocity is zero, that is, the waves do not propagate.

and remembering that $K_p = V_p''$ (3.4),

$$\omega^2 = \frac{2}{M} \sum_{p>0} V_p''(1 - \cos p\mathbf{k} \cdot \mathbf{a})$$ (3.9)

This is the dispersion relation for the infinite crystal. We will discuss it, without loss of generality, in the simple case where the interaction is limited to the nearest neighbor planes $(p=1)$. We then have:

$$\omega^2 = \frac{2}{M} V''(1 - \cos \mathbf{k} \cdot \mathbf{a}) = \frac{4}{M} V'' \sin^2\left(\frac{\mathbf{k} \cdot \mathbf{a}}{2}\right)$$

or

$$\omega = 2\left(\frac{V''}{M}\right)^{1/2} \left|\sin\left(\frac{\mathbf{k} \cdot \mathbf{a}}{2}\right)\right|$$ (3.10)

The dispersion curve is given in Figure 3.2.

We can make the following observations:

i. $\omega = f(\mathbf{k})$ is periodic with the period $|\mathbf{g}| = |2\pi/\mathbf{a}|$, equal by definition to the period of the reciprocal lattice. The interval $[-\pi/a, +\pi/a]$ defines the first *Brillouin zone*.

ii. $\omega = f(\mathbf{k})$ is a symmetric function. It is therefore sufficient to specify it in the interval $[0, \pi/a]$.

iii. At the edge of the Brillouin zone, that is, for $|\mathbf{k}| = (2n+1)\pi/a = (n+1/2)|\mathbf{g}|$, the frequency is a maximum:

$$\omega = \omega_{max} = 2\left(\frac{V''}{M}\right)^{1/2} \tag{3.11}$$

(The maximum atomic vibrational frequency is of the order of 10^{13} Hz.)

At the maximum, $d\omega/dk = 0$, which means that the group velocity of the waves vanishes. The only permissible wave is a stationary wave that does not propagate energy:

$$u = u^0 \exp i[(n + \tfrac{1}{2})\mathbf{g} \cdot \mathbf{a} - \omega t] = \pm u^0 \exp(-i\omega t) \tag{3.12}$$

In the long wavelength limit the neighboring planes vibrate almost in phase and we can write:

$$\omega \cong ak\left(\frac{V''}{M}\right)^{1/2} \tag{3.13}$$

The frequency is proportional to the wave number. In other words, the group velocity $d\omega/dk$ is equal to the phase velocity: there is no dispersion. The phase velocity is equal to the velocity v_p of the longitudinal wave:

$$\frac{d\omega}{dk} = a\left(\frac{V''}{M}\right)^{1/2} = \frac{\omega}{k} = v_P \tag{3.14}$$

Indeed, if the wave vector k is much smaller than the reciprocal lattice parameter, that is, if the lattice parameter in real space is much smaller than the wavelength, it is reasonable to approximate the lattice as an elastic continuum in which the wave equation for longitudinal waves is (2.20) and the velocity of the waves is given by (2.14): $v_P = [(\lambda + 2\mu)/\rho]^{1/2}$.

The reasoning would, of course, be the same for transverse waves. Indeed, there are three dispersion curves, one for the P waves and one for each polarization of the S waves.

Comparing the classic expression (2.14) for the velocity of the P waves with (3.14), we see that the relevant elastic modulus is proportional to the second derivative of the potential energy with respect to strain. Therefore, the elastic constants, introduced as phenomenological material constants in the thermodynamic approach, can be physically interpreted in terms of interatomic potentials.

As an example, let us calculate the value of the bulk modulus K_0 at 0 K (no thermal energy), for a simple ionic crystal such as NaCl (Kittel 1967). The pair interaction potential E_{ij} between neighboring ions of opposite charge consists of an attractive Coulombic part and a short-range repulsive part due to the ion cores:

$$E_{ij} = \frac{q_i q_j}{r_{ij}} + \frac{ZB}{r_{ij}^n} \tag{3.15}$$

q_i and q_j are the electric charges of ion i and its neighbor j, r_{ij} is the distance between the ions, and B and n are parameters of the repulsive part of the potential (Born potential).

The *cohesive energy* E_c of the crystal is obtained by summing the attractive parts of the potential (a somewhat complicated process) and assuming that the short-range repulsive part extends only to nearest neighbors:

$$E_c = N\left(-\frac{\alpha q^2}{R} + \frac{ZB}{R^n}\right) \tag{3.16}$$

where $\alpha \equiv \Sigma_j(\pm)R/r_{ij}$ is the *Madelung constant* ($\alpha \cong 1.75$ for the NaCl structure), R is the nearest neighbor distance, $q = q_i = -q_j$ is the electric charge, Z is the coordination number, and N is the total number of ions of one sign.

At equilibrium, the nearest neighbor distance is R_0, given by $(dE_c/dR)_{R=R_0} = 0$, and we have then:

$$R_0^{n-1} = \frac{nZB}{\alpha q^2} \tag{3.17}$$

Hence

$$E_c = -\frac{N\alpha q^2}{R_0}\left[\frac{R_0}{R} - \frac{1}{n}\left(\frac{R_0}{R}\right)^n\right] \tag{3.18}$$

We can now calculate $K_0 = -V(dP/dV)_{R=R_0}$.

From $P = -(\partial E/\partial V)_S$, we get $(\partial P/\partial V)_S = -(\partial^2 E/\partial V^2)_S$ and $K_0 = V d^2E/dV^2$. For NaCl structure, we have $N = 4$ formular units per face centered cubic unit cell, each occupying a volume $a^3/4$ ($a = 2R$ is the cell parameter), hence $V = 2NR^3$.

$$\frac{dE_c}{dV} = \frac{dE_c}{dR}\frac{dR}{dV} = \frac{dE}{dR}\frac{1}{6NR^2}$$

$$\frac{d^2E_c}{dV^2} = \frac{d^2E_c}{dR^2}\left(\frac{dR}{dV}\right)^2 + \frac{dE_c}{dR}\frac{d^2R}{dV^2}$$

At equilibrium, $dE/dR = 0$ and $R = R_0$, hence

$$K_0 = V\left(\frac{1}{6NR^2}\right)^2\left(\frac{d^2E_c}{dR^2}\right)_{R=R_0} = \frac{1}{18NR_0}\left(\frac{d^2E_c}{dR^2}\right)_{R=R_0} \tag{3.19}$$

From (3.16), (3.17), and (3.19) we obtain

$$K_0 = \frac{1}{R_0^3}\frac{\alpha q^2(n-1)}{18R_0} \tag{3.20}$$

In the general case of an infinite crystal, with p atoms per primitive unit cell (not necessarily of the same chemical nature), it can be shown (see Kittel 1967) that for each value of the wave number k, there are $3p$ frequencies, each corresponding to one normal mode. There are:

Three acoustical modes corresponding to one longitudinal (P) mode and two transverse (S) modes, if they are pure. The modes are orthogonal to one another.

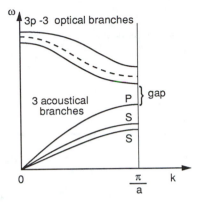

Figure 3.3. Dispersion curve (half of the first Brillouin zone) of an infinite lattice with p atoms per primitive unit cell. There are three acoustical modes (1 P mode and 2 S modes polarized at right angles) and $3p-3$ optical modes. There is a gap of forbidden energies at the edge of the Brillouin zone.

$3p-3$ *optical modes* corresponding to out-of-phase vibrations of neighboring planes for small wave numbers. The optical modes often have frequencies in the range of that of infrared or visible light and can cause optical absorption, hence their name. Near the edge of the Brillouin zone, there is a forbidden gap (Fig. 3.3), electromagnetic waves with a frequency within the gap cannot propagate in the crystal and are reflected.

3.2.2 Density of states of a finite lattice

In the case of a finite lattice, the number of degrees of freedom, hence, of possible normal modes, is finite and the vibrations are quantized: Instead of continuous dispersion curves, we have a succession of discrete points, one for every one of the allowed wave numbers (Fig. 3.4).

We also need boundary conditions: For a crystal large enough compared with the interatomic distance a, which boundary conditions are chosen does not matter much as long as there are boundary conditions. The ones currently used are the Born–von Karman periodic boundary conditions: For N parallel planes in the crystal, we impose the condition that the vibrational state of the Nth (last) plane be the same as that of the first, which amounts to ideally closing the crystal on itself (as a hypertorus in 4-D space), hence imposing a period N:

$$u_{n+N} = u_n$$

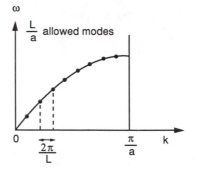

Figure 3.4. Dispersion "curve" of a finite unidimensional lattice of length L and period a; it consists of L/a discrete points, one for each allowed mode.

We have therefore

$$u^0 \exp i[n\mathbf{k}\cdot\mathbf{a} - \omega t] = u^0 \exp i[(n+N)\mathbf{k}\cdot\mathbf{a} - \omega t]$$

Hence

$$\exp iN\mathbf{k}\cdot\mathbf{a} = 1$$

or

$$N\mathbf{k}\cdot\mathbf{a} = 2m\pi$$

with m a positive or negative integer.

The length of the crystal is $L = Na$. The allowed modes therefore have wave numbers given by

$$k = \frac{2m\pi}{L} \tag{3.21}$$

and the number of allowed wave numbers is

$$\frac{2\pi}{a}\left(\frac{2\pi}{L}\right)^{-1} = \frac{L}{a} = N \tag{3.22}$$

Let us now generalize this result to the three-dimensional reciprocal space (k-space) and consider, for the sake of simplicity, the case of a crystal of volume V and primitive unit cell volume V_L.

The volume of the Brillouin zone is $(2\pi)^3/V_L$, and the volume per allowed wave number in k-space is $(2\pi)^3/V$. There are therefore V/V_L allowed values of k in the Brillouin zone.

The *density of states* $g(\omega)$ is the number of modes per unit frequency range. The number of vibrational states between ω and $\omega + d\omega$ is therefore:

$$g(\omega)\,d\omega = w(k)\,dk \tag{3.23}$$

Table 3.1. *Volume and mass of the various subunits in a polyatomic crystal*

Unit	Volume	Mass	Number of subunits
Mole	V	ρV	Unit cells: V/V_L Formulas: $VZ/V_L = N_A$ Atoms: $nN_A = VZn/V_L$
Unit cell	V_L	ρV_L	Formulas: Z Atoms: Zn
Formula ("molecule")	$V_L/Z = V/N_A$	$\rho V_L/Z = M$	Atoms: n
Atoms	V/nN_A	$M/n = \bar{M}$	

where $w(k)dk$ is the number of states in k-space in a spherical shell of thickness dk, between k and $k+dk$. If we take the volume of the crystal equal to the molar volume V, and if the crystal has n atoms per formula unit and Z unit cells per mole, then there are nN_A atoms in the mole (N_A is Avogadro's number) (see Table 3.1) and $3nN_A$ modes in the volume of the Brillouin zone (see Kieffer 1979a). We have

$$w(k)dk = \frac{3nN_A V_L}{(2\pi)^3} \cdot 4\pi k^2 dk = \frac{3nZV}{(2\pi)^3} \cdot 4\pi k^2 dk$$

We will now calculate the density of states in the case of the very useful Debye approximation.

3.3 Debye's approximation

3.3.1 Debye frequency

Debye's approximation consists in assuming that the long wave or continuum approximation, with a linear dispersion curve $\omega = v_m k$ (where v_m is an average velocity of sound waves), holds for the whole vibrational spectrum. In other words, all the modes are considered to be acoustic, with the same average value of the velocity. The allowed ks are assumed to be uniformly distributed in the Brillouin zone. If we take for the volume V of the crystal the molar volume, the crystal contains nN_A atoms, where n is the number of atoms in the formular unit and N_A is Avogadro's number.

The density of states (3.23) is therefore given by

$$g(\omega) = \frac{3nZV}{(2\pi)^3} \cdot 4\pi \frac{\omega^2}{v_m^3}$$

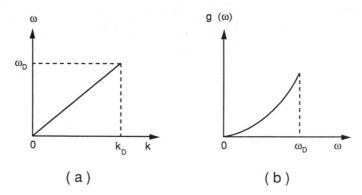

Figure 3.5. Dispersion curve (a) and vibrational spectrum (b) in the case of the Debye approximation. The vibrational spectrum is the curve of the density of states $g(\omega)$ vs. frequency. ω_D and k_D are the Debye temperature and wave number, respectively.

or

$$g(\omega) = A\omega^2 \tag{3.24}$$

with

$$A = \frac{3nZV}{2\pi^2 v_m^3} \tag{3.25}$$

The curve of the density of states versus frequency is the vibrational spectrum; we see that in Debye's approximation it is parabolic (Fig. 3.5).

Debye's calculation of the average sound velocity gives

$$\blacktriangleright \qquad v_m = 3^{1/3}\left(\frac{1}{v_P^3} + \frac{2}{v_S^3}\right)^{-1/3} \tag{3.26}$$

O. L. Anderson (1963) showed that the Debye average sound velocity is accurately estimated by using the Voigt–Reuss–Hill averaging method (see Section 2.2) for calculating the longitudinal and transverse sound velocities v_P and v_S from single-crystal elastic constants.

The Brillouin zone is assumed to have the simple shape of a sphere with radius k_D given by

$$\frac{4}{3}\pi(k_D)^3 = \frac{(2\pi)^3}{V_L}$$

The maximum radius k_D corresponds to a maximum cut-off frequency ω_D, called the *Debye frequency:* $\omega_D = k_D v_m$.

$$\omega_D = v_m\left(\frac{6\pi^2}{V_L}\right)^{1/3} = v_m\left(\frac{6\pi^2 N_A}{Z}\right)^{1/3} V^{-1/3} \tag{3.27}$$

However, if we assume that all the different atoms play equivalent mechanical roles in the vibrations, it is then possible to consider the individual atom (of any chemical nature) as the vibrational unit. In that case, we can take for V_L the average volume of one atom:

$$(V_L)^{-1} = \frac{nN_A}{V} = \frac{nN_A \rho}{M}$$

where M is the mass of the formula unit. Hence

$$\omega_D = v_m (6\pi^2 nN_A)^{1/3} V^{-1/3} \tag{3.28}$$

Introducing the mean atomic mass $\bar{M} = M/n$, we find the expression of Debye's frequency in general use (O. L. Anderson 1988; Robie and Edwards 1966):

$$\omega_D = (6\pi^2 N_A)^{1/3} \left(\frac{\rho}{\bar{M}}\right)^{1/3} v_m \tag{3.29}$$

Since this approach assumes that there is only one atom per unit cell, it follows that all the modes are assumed to be acoustic, which is consistent with Debye's approximation.

3.3.2 Vibrational energy and Debye temperature

In the limit of the linear continuum approximation, the normal vibration modes are independent, hence the energy of one lattice mode (state) depends only on its frequency ω and on the number of quanta of vibration occupying that state (phonon occupancy). In thermal equilibrium, the phonon occupancy is given by the Bose–Einstein distribution (phonons are bosons):

$$\langle n \rangle = \left[\exp\left(\frac{\hbar\omega}{k_B T}\right) - 1\right]^{-1} \tag{3.30}$$

where $\hbar = h/2\pi = 1.0546 \times 10^{-34}$ Js ($h = 6.6262 \times 10^{-34}$ Js is Planck's constant) and $k_B = 1.3807 \times 10^{-23}$ JK^{-1} is Boltzmann's constant.

The crystal, with all its modes of oscillation, can be considered as a collection of oscillators. *Einstein's approximation* assumes that all the oscillators have the same frequency; neglecting the zero-point energy, the energy of the crystal with N oscillators is therefore

$$E = 3N\langle n \rangle \hbar\omega = 3N\hbar\omega \left[\exp\left(\frac{\hbar\omega}{k_B T}\right) - 1\right]^{-1}$$

and the specific heat is

$$C_V = \left(\frac{\partial E}{\partial T}\right)_V = 3Nk_B \left(\frac{\hbar\omega}{k_B T}\right)^2 \left(\exp\frac{\hbar\omega}{k_B T}\right)\left(\exp\frac{\hbar\omega}{k_B T} - 1\right)^{-2}$$

In *Debye's approximation*, we have various permissible frequencies: $\omega(k) \leq \omega_D$. The energy per oscillator (per mode) is equal to $\hbar\omega$; since we have $g(\omega)\,d\omega$ modes in the frequency range ω to $\omega+d\omega$, the energy (neglecting the zero-point energy) is

$$E = \int_0^{\omega_D} \langle n(\omega, T)\rangle \hbar\omega g(\omega)\, d\omega \tag{3.31}$$

With (3.24), (3.25), and (3.30) we have

$$E = \frac{3\hbar nZV}{2\pi^2 v_m^3} \int_0^{\omega_D} \frac{\omega^3 d\omega}{\exp(\hbar\omega/k_B T)-1} \tag{3.32}$$

It is convenient to change variables and express (3.32) in terms of the nondimensional variables $x = \hbar\omega/k_B T$ and $x_D = \Theta_D/T$, where Θ_D is the *elastic Debye temperature,* defined by

$$\Theta_D = \frac{\hbar}{k_B}\omega_D \tag{3.33}$$

With (3.27)

$$\Theta_D = \frac{\hbar}{k_B}\left(\frac{6\pi^2 N_A}{Z}\right)^{1/3} V^{-1/3} v_m = \frac{h}{k_B}\left(\frac{3N_A}{4\pi Z}\right)^{1/3} V^{-1/3} v_m \tag{3.34}$$

which, if $Z=1$, gives

$$\Theta_D = 251.2 V^{-1/3} v_m \tag{3.35}$$

with V in cm³/mole and v_m in km/s given by (3.26).

With (3.28) and (3.29), that is, if all the atoms play an equivalent mechanical role,

$$\Theta_D = \frac{h}{k_B}\left(\frac{3nN_A}{4\pi}\right)^{1/3} V^{-1/3} v_m = \frac{h}{k_B}\left(\frac{3N_A}{4\pi}\right)^{1/3}\left(\frac{\rho}{\overline{M}}\right)^{1/3} v_m \tag{3.36}$$

or

$$\blacktriangleright \qquad \Theta_D = 251.2\left(\frac{\rho}{\overline{M}}\right)^{1/3} v_m \tag{3.37}$$

There are various more or less sophisticated ways of calculating Debye's temperature from elastic constants (Alers 1965) or from specific heats (Blackman 1955) and the resulting numerical values are often quite different. When comparing and using Debye temperatures of various materials found in the literature, it is always advisable to check whether they have been calculated in the same way (e.g., using (3.35) or (3.37) (see Table 3.2).

Table 3.2. *Physical constants of some typical and important crystals*

Material	ρ	$\bar{M}$	K_S	μ	v_P	v_S	v_m	Θ_D	T_m
Iron[1] Fe	7.87	55.85	172.7	83.1	6.00	3.25	3.63	474	1808
Nickel[1] Ni	8.91	58.69	185.6	88.2	5.83	3.15	3.51	471	1726
Lead[1] Pb	11.34	207.20	39.9	9.0	2.14	0.89	1.00	96	600
Copper[1] Cu	8.92	63.55	144.9	49.4	4.86	2.35	2.64	345	1356
Diamond[1] C	3.51	12.01	584.8	346.3	17.27	9.93	11.03	1839	3823
Silicon[1] Si	2.33	28.09	97.9	66.7	8.95	5.35	5.92	649	1683
Halite[2] NaCl	2.16	29.22	24.7	14.4	4.51	2.58	2.87	302	1075
Periclase[2] MgO	3.58	20.15	162.8	129.4	9.68	6.01	6.63	936	3125
Stishovite[2] SiO₂	4.29	20.03	277.4	232.2	11.70	7.36	8.10	1217	
Corundum[2] Al₂O₃	3.99	20.39	251.9	162.0	10.83	6.37	7.06	1030	2345
Forsterite[2] Mg₂SiO₄	3.21	20.10	128.2	80.5	8.57	5.00	5.55	757	2183
β-phase[3] Mg₂SiO₄	3.47	20.10	174.0	114.0	9.69	5.73	6.35	888	
Spinel[4] Mg₂SiO₄	3.56	20.10	184.0	119.0	9.81	5.78	6.41	904	
Pyrope[2] Mg₃Al₂Si₃O₁₂	3.56	20.16	176.6	89.6	9.12	5.02	5.59	788	
Enstatite[2] MgSiO₃	3.20	20.08	107.5	75.4	8.06	4.85	5.37	731	
Ilmenite[5] MgSiO₃	3.80	20.08	212.0	132.0	10.10	5.89	6.54	943	
Perovskite[6] MgSiO₃	4.11	20.08	246.4	184.2	10.94	6.69	7.39	1094	
Perovskite[2] CaTiO₃	4.04	27.19	177.0	104.0	8.84	5.07	5.64	750	2248

Notes: Specific mass ρ in g/cm³, mean atomic mass $\bar{M}$ in g/at. Melting temperature T_m and Debye temperature Θ_D calculated from (3.37), in Kelvins. Voigt–Reuss–Hill, at room temperature and ambient pressure, bulk modulus K_S and shear modulus μ in GPa.
Sources: [1]Simmons and Wang (1971); [2]Sumino and Anderson (1984); [3]Sawamoto et al. (1984); [4]Weidner et al. (1984); [5]Weidner and Ito (1985); [6]Yeganeh-Haeri, Weidner, and Ito (1988).

Using (3.34) we have

$$E = \frac{3nZVk_B^4 T^4}{2\pi^2 \hbar^3 v_m^3} \int_0^{x_D} \frac{x^3}{\exp x - 1} \, dx \qquad (3.38)$$

or

$$E = 9nN_A k_B T x_D^{-3} \int_0^{x_D} \frac{x^3}{\exp x - 1} \, dx \qquad (3.39)$$

3.3.3 Specific heat

The specific heat or heat capacity at constant volume C_V is obtained by differentiating (3.39) with respect to temperature:

$$C_V = \left(\frac{dE}{dT} \right)_V = 9nN_A k_B x^{-3} \int_0^{x_D} \frac{x^4 \exp x}{(\exp x - 1)^2} \, dx \qquad (3.40)$$

or

$$C_V = 9nN_A k_B \mathfrak{D} \left(\frac{\Theta_D}{T} \right) \qquad (3.41)$$

The *Debye function* $\mathfrak{D}(\Theta_D/T)$ is calculated and tabulated (e.g., in Landolt–Börnstein tables).

At $T \gg \Theta_D$ the heat capacity approaches the classical value given by the Dulong and Petit law:

$$C_V = 3nNk_B = 3nR \qquad (3.42)$$

with the gas constant $R \cong 2 \, \text{cal/mol K}$.

At low temperatures, the heat capacity is approximately equal to

$$C_V \cong 234nNk_B \left(\frac{T}{\Theta_D} \right)^3 \qquad (3.43)$$

(Debye's T^3 law).

The values of $C_V(T)$ (Fig. 3.6) can be experimentally determined by calorimetry and fitted to equation (3.41) by choosing the best value of the Debye temperature. The Debye temperature determined in this fashion is called the *calorimetric Debye temperature* (at temperature T).

3.3.4 Validity of Debye's approximation

Debye's approximation is only as good as its basic assumptions. It is therefore valid if the actual vibrational spectrum can reasonably be approximated by a parabolic curve $g(\omega)$ and if most of the spectrum corresponds to frequencies lower than the cut-off Debye frequency ω_D. If Debye's

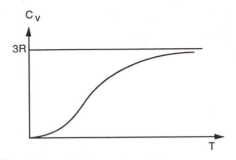

Figure 3.6. Typical curve of the specific heat at constant volume C_V vs. temperature T. At low temperatures, C_V varies as T^3 and at high temperatures C_V approaches the classical Dulong and Petit value $3R$.

model is valid, then the calorimetric Debye temperature must be independent of temperature and equal to the elastic Debye temperature (usually calculated with (3.37) and assuming reasonably that the elastic constants, varying little with temperature, can be taken equal to their values at room temperature). This is practically the case for elements and simple close-packed substances, for which it can be reasonably assumed that all atoms are mechanically equivalent (Kieffer 1979a) (Fig. 3.7). Plots of the ratio $\Theta_{Dcal}/\Theta_{Del}$ as a function of temperature for minerals with open structures (Fig. 3.8) exhibit a dip at low temperatures corresponding to the actual presence of more low-frequency modes (optical modes) than taken into account in Debye's model and rise steadily at high temperatures, indicating an excess of modes near Debye's frequency and a deficit at higher frequencies in the model (Kieffer 1979a). O. L. Anderson (1988) finds that Debye's model is satisfactory for close-packed minerals with a volume per atom $V_a = \bar{M}/\rho$ smaller than about 5.8 cm^3/mol, such as stishovite, corundum, and periclase (Fig. 3.9). It is reasonable to think that Debye's model is probably valid for the dense lower mantle perovskite form of $MgSiO_3$ ($V_a = 4.8$).

Kieffer (1979) developed a generalized model based on lattice dynamics and observational vibrational data (infrared and Raman spectroscopy, inelastic neutron scattering) that allows a better calculation of thermodynamic data of minerals (specific heat and entropy) than Debye's model.

3.4 Mie–Grüneisen equation of state

We have calculated, in Debye's approximation, the vibrational (i.e., thermal) energy of a solid. We are now in a position to calculate the total

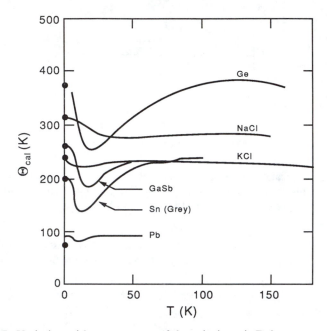

Figure 3.7. Variation with temperature of the calorimetric Debye temperature for a few simple substances. Elastic Debye temperatures at 0 K are shown by closed circles (after Kieffer 1979).

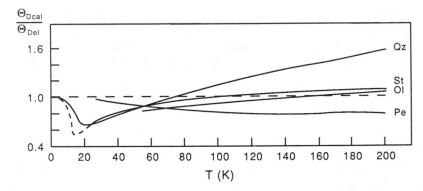

Figure 3.8. Variation with temperature of the ratio of the calorimetric to the elastic Debye temperatures for quartz (Qz), olivine (Ol), stishovite (St), and periclase (Pe) (after Kieffer 1979).

energy of the solid and its equation of state, that is, the expression of pressure P as a function of specific volume V and temperature T.

Since $P = -(\partial F/\partial V)_T$, let us start with Helmholtz's free energy F and consider it as the sum of two terms

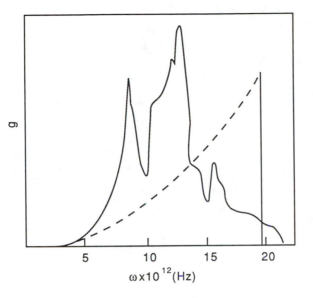

Figure 3.9. Vibrational spectrum (density of states) for periclase compared with the Debye spectrum (dashed curve). The frequencies of the modes, including the optical modes, are nearly all below the Debye frequency and both spectra converge at low frequencies (after O. L. Anderson 1988).

$$F = E_C + F_D \qquad (3.44)$$

where E_C is the cohesive energy at 0 K, calculated in Section 3.2 (3.16), and F_D is the free energy of the phonon "gas" in Debye's approximation

$$F_D = E_D - TS_D = E_D + T\left(\frac{\partial F_D}{\partial T}\right)_V \qquad (3.45)$$

E_D is given by (3.39) and we will write it

$$E_D = Tf\left(\frac{\Theta_D}{T}\right)$$

Now, from (3.45) we have

$$E_D = F_D - T\left(\frac{\partial F_D}{\partial T}\right)_V = \frac{d(F_D/T)}{d(1/T)} \qquad (3.46)$$

It will be convenient to express F_D also as a function of T and Θ_D/T

$$F_D = Tg\left(\frac{\Theta_D}{T}\right) \qquad (3.47)$$

which, with (3.46), gives

$$E_D = \Theta_D \frac{dg(\Theta_D/T)}{d(\Theta_D/T)} \tag{3.48}$$

Let us now write the expression of pressure:

$$P = -\left(\frac{\partial F}{\partial V}\right)_T = -\frac{dE_C}{dV} - \frac{dF_D}{d\Theta_D}\frac{d\Theta_D}{dV} \tag{3.49}$$

Now,

$$\frac{dF_D}{d\Theta_D} = \frac{dF_D}{d(\Theta_D/T)}\frac{d(\Theta_D/T)}{d\Theta_D} = \frac{1}{T}\frac{d(Tg(\Theta_D/T))}{d(\Theta_D/T)}$$

and, with (3.48),

$$\frac{dF_D}{d\Theta_D} = \frac{E_D}{\Theta_D} \tag{3.50}$$

We can therefore write

$$P = -\frac{dE_C}{dV} - \frac{E_D}{\Theta_D}\frac{d\Theta_D}{dV} \tag{3.51}$$

We have thus expressed P in terms of known quantities and of the variation of Debye's temperature with volume. It is convenient to introduce the *Debye–Grüneisen parameter*

$$\blacktriangleright \qquad\qquad \gamma_D = -\frac{d\ln\Theta_D}{d\ln V} = -\frac{d\ln\omega_D}{d\ln V} \tag{3.52}$$

We have then

$$\blacktriangleright \qquad\qquad P = -\frac{dE_C}{dV} + \gamma_D\frac{E_D}{V} \tag{3.53}$$

This is the complete Mie–Grüneisen equation of state, which we first encountered in Section 2.3.3. The thermal pressure was then related to the increase in thermal energy by the thermodynamic Grüneisen parameter γ_{th}. Now we see that the thermal energy is the energy of the phonon gas and we have introduced a more physical definition of the Grüneisen parameter, expressing the volume dependence of the Debye frequency.

We will now show that the two definitions are equivalent and that $\gamma_{th} = \gamma_D$.

Assuming that the mode frequencies are independent of temperature (quasi-harmonic approximation, see Sec. 3.6.2), we can differentiate (3.53) with respect to T and obtain

$$\left(\frac{\partial P}{\partial T}\right)_V = \frac{\gamma_D}{V}\left(\frac{\partial E_D}{\partial T}\right)_V = \frac{\gamma_D}{V}C_V$$

Now, by definition of the coefficient of thermal expansion α and equation (1.27), we have

$$\alpha = \frac{1}{V}\left(\frac{\partial V}{\partial T}\right)_P = -\frac{1}{V}\left(\frac{\partial V}{\partial P}\right)_T\left(\frac{\partial P}{\partial T}\right)_V = \frac{1}{K_T}\left(\frac{\partial P}{\partial T}\right)_V \qquad (3.54)$$

Hence

$$\gamma_D = \frac{\alpha V K_T}{C_V} = \gamma_{th} \qquad (3.55)$$

3.5 The Grüneisen parameters

So far we have introduced two definitions of the Grüneisen parameter, a thermodynamic, that is, macroscopic, one γ_{th}, and a microscopic one γ_D. Now, there are many more ways of defining a Grüneisen parameter or "gamma" and we will review here the more important and currently encountered ones in the literature (see Stacey 1977a, b).

We should indeed start with Grüneisen's own definition: He considered what we call the "*mode gamma*" γ_i expressing the volume dependence of the frequency of the ith vibration mode of the lattice

$$\gamma_i = -\frac{\partial \ln \omega_i}{\partial \ln V} = \frac{\partial \ln \omega_i}{\partial \ln \rho} \qquad (3.56)$$

If all the ω_is have the same volume dependence and there is only one Debye cut-off frequency, all the mode gammas are equal to the *Debye gamma* γ_D. If, in addition we make the assumption that the mode frequencies are independent of temperature, it can be shown to be equivalent to the thermodynamic gamma (see Section 3.4).

Slater (1939) asserted that Debye's theory implied that all mode frequencies varied in the same ratio as the cut-off frequency, hence that all the mode gammas were equal. He also made the assumption that Poisson's ratio ν was independent of the volume and derived another expression for Grüneisen's parameter in terms of the variation of the bulk modulus K with volume as follows:

From (3.28) we have

$$\omega_D \propto v_m V^{-1/3} \propto v_m \rho^{1/3} \qquad (3.57)$$

We can show, from (2.12), (2.14), and (2.15) that the velocities of the longitudinal and transverse waves can be expressed as the square root of K/ρ times a function of Poisson's ratio:

$$v_P = \left(\frac{K}{\rho}\right)^{1/2}\left[\frac{3(1-\nu)}{1+\nu}\right]^{1/2}$$

$$v_S = \left(\frac{K}{\rho}\right)^{1/2}\left[\frac{3(1-2\nu)}{2(1+\nu)}\right]^{1/2}$$

(3.58)

N.B. We clearly see here that the assumption that Poisson's ratio is independent of volume is in fact already included in Debye's model stricto sensu, since a dependence of ν on volume would mean that the longitudinal and transverse velocities, hence the frequencies of the corresponding modes, would have different volume dependences. The modes would not all have the same gammas, in contradiction with Debye's assumptions.

The average Debye velocity can therefore be written

$$v_m = K^{1/2}V^{1/2}f(\nu)$$

(3.59)

It follows that

$$\ln \omega_D = \tfrac{1}{2}\ln K + \tfrac{1}{6}\ln V + \ln f(\nu)$$

(3.60)

and, since ν is assumed to be independent of volume,

$$\gamma_S = -\frac{d\ln \omega_D}{d\ln V} = -\frac{1}{6} - \frac{1}{2}\frac{d\ln K}{d\ln V}$$

(3.61)

or

$$\gamma_S = -\frac{1}{6} + \frac{1}{2}\frac{d\ln K}{d\ln \rho}$$

(3.62)

or

▶
$$\gamma_S = -\frac{1}{6} + \frac{1}{2}\frac{dK}{dP}$$

(3.63)

This defines the *Slater gamma,* which explicitly depends on the assumption that Poisson's ratio is independent of volume (or pressure). Using the definition of the bulk modulus

$$K_T = -V\left(\frac{\partial P}{\partial V}\right)_T$$

we can write

$$\gamma_S = -\frac{1}{6} - \frac{1}{2}\left[1 + \frac{\partial \ln(-\partial P/\partial V)_T}{\partial \ln V}\right]$$

which gives another expression of Slater's gamma

$$\gamma_S = -\frac{2}{3} - \frac{V}{2}\frac{(\partial^2 P/\partial V^2)_T}{(\partial P/\partial V)_T}$$

(3.64)

Conversely, knowing Slater's gamma, we can obtain the variation with density or pressure of the bulk modulus:

$$\frac{d\ln K}{d\ln\rho} = \frac{dK}{dP} = 2\gamma_S + \frac{1}{3} \qquad (3.65)$$

Since the ratio K/μ depends only on Poisson's ratio, which is assumed not to depend on volume, we also have:

$$\frac{d\ln\mu}{d\ln\rho} = 2\gamma_S + \frac{1}{3} \qquad (3.66)$$

and

$$\frac{d\mu}{dP} = \frac{\mu}{K}\left(2\gamma_S + \frac{1}{3}\right) \qquad (3.67)$$

The assumption that Poisson's ratio does not vary as specific volume decreases is of course not strictly correct, as Slater perfectly knew: Poisson's ratio does, in fact, slightly increase with pressure. Slater's gamma is nevertheless a rather good approximation for close-packed solids (O. L. Anderson 1988) and formulas (3.64)–(3.67) are useful expressions of the pressure dependence of the elastic moduli.

The restriction that Poisson's ratio be independent of volume can be lifted by dealing separately with the longitudinal and transverse acoustic branches of the phonon spectrum and defining a transverse gamma γ_t and a longitudinal gamma γ_l; this, of course, is in contradiction with the original Debye assumptions and implies a modification of Debye's model (Stacey 1977a, b). The weighted average of γ_t and γ_l defines a new gamma called the *acoustic gamma* or *high-temperature gamma* (Vashchenko and Zubarev 1963; Stacey 1977a, b):

$$\gamma_a = \gamma_{HT} = \tfrac{1}{3}(\gamma_l + 2\gamma_t) \qquad (3.68)$$

Defining γ_l and γ_t in the same way as γ_D, but replacing v_m by v_P and v_S, respectively, and using (3.58), we obtain:

$$\gamma_a = \frac{1}{2}\frac{d\ln K}{d\ln\rho} - \frac{1}{6} + \frac{1}{3}\frac{5v^2-4v}{(1-v^2)(1-2v)}\frac{d\ln v}{d\ln\rho} \qquad (3.69)$$

We see that the acoustic gamma is equal to the Slater gamma plus a term which is a function only of Poisson's ratio and its dependence on specific mass. Knopoff and Shapiro (1969) show that in the acoustic gamma the pressure derivative of the shear modulus plays a role significantly more important than the pressure derivative of the bulk modulus.

The acoustic gamma is also called the high-temperature gamma because it can be derived from the expression for thermal pressure, assuming that at high temperature the thermal energy $E_{th} = 3RT$ is equally distributed between the P-mode and the two S-modes. We have then

$$P_{th} = \frac{RT}{V}(\gamma_l + 2\gamma_t) = \frac{3RT}{V}\gamma_{HT}$$

and the definition (3.68) immediately follows.

Quareni and Mulargia (1988) propose to use what they call the *Debye–Brillouin gamma*

$$\gamma_{Br} = \frac{1}{3} - V\left(\frac{dv_P}{dV} + 2\frac{dv_S}{dV}\right)(2v_S + v_P)^{-1} \tag{3.70}$$

To compare it with the other formulations, it is convenient to write

$$\gamma_{Br} = \frac{1}{3} + \frac{d\ln[(v_P + 2v_S)/3]}{d\ln\rho} \tag{3.71}$$

Note that we could write the acoustic gamma as

$$\gamma_a = \frac{1}{3} + \frac{1}{3}\left(\frac{d\ln v_P}{d\ln\rho} + 2\frac{d\ln v_S}{d\ln\rho}\right) = \frac{1}{3} + \frac{d\ln(v_P v_S^2/3)}{d\ln\rho} \tag{3.72}$$

and, of course, Debye's gamma is

$$\gamma_D = \frac{1}{3} + \frac{d\ln[(1/v_P^3 + 2/v_S^3)/3]^{-1/3}}{d\ln\rho} \tag{3.73}$$

We see that in practice these formulations differ only in the way the velocities are averaged; they can all be written

$$\gamma = \frac{1}{3} + \frac{d\ln\bar{v}}{d\ln\rho} \tag{3.74}$$

with

$$\bar{v}_D = 3^{1/3}\left(\frac{1}{v_P^3} + \frac{2}{v_S^3}\right)^{-1/3} \quad \text{for the Debye gamma}$$

$$\bar{v}_a = (v_P v_S^2)^{1/3} \qquad \text{for the acoustic gamma} \tag{3.75}$$

$$\bar{v}_{Br} = \tfrac{1}{3}(v_P + 2v_S) \qquad \text{for the Debye–Brillouin gamma}$$

The pressure dependence of Poisson's ratio is obviously included in the last two expressions and the cut-off frequencies for the longitudinal and transverse branches may or may not be assumed equal without affecting their validity.

All these approaches use the same fundamental assumptions of the Debye approximation: The optical modes are not specifically taken into account and all the modes are assumed to be independent and nondispersive (i.e., the group velocity is equal to the phase velocity and does not depend on the frequency), an approximation valid only at low frequencies and for acoustic modes only. At high temperatures, the high-frequency modes carry an important proportion of the thermal energy and they are dispersive. Furthermore, they interact and cannot be considered as independent.

Irvine and Stacey (1975) proposed another way of calculating an expression for the Grüneisen parameter in the high-temperature classical limit, still as a function of the pressure derivative of the bulk modulus, but avoiding consideration of lattice modes (see also Stacey 1977a, b). The expression they found is identical with the one derived earlier by Vashchenko and Zubarev (1963) from the free-volume theory. In view of the fact that the *Vashchenko–Zubarev gamma* or *free-volume gamma* is widely quoted and used, we will first give the original demonstration before giving Irvine and Stacey's demonstration.

The free-volume theory was proposed by Lennard-Jones and Devonshire (1937) to account for the properties of dense gases and liquids. The main idea is that the atoms can be regarded as confined in cells and that their environment is approximately the same as in a crystal. The atoms are in potential wells resulting from the attractive and repulsive parts of the pair interaction potentials with their neighbors. This is in a way similar to the Einstein approximation where the atoms are considered as independent oscillators in harmonic, parabolic, potential wells, except that here the wells do not have to be parabolic. (Interestingly enough Vashchenko and Zubarev used for a solid a theory that was devised for dense fluids, assuming they could be locally regarded as solid-like!) Due to the fact that the energy is high near the cell boundaries, the atoms are in effect confined to a "free volume" V_F smaller than the cell size

$$V_F = \int_{\text{cell}} \exp\left[-\frac{\chi(r) - \chi(0)}{kT}\right] dV \tag{3.76}$$

where $\chi(r)$ is the potential energy of the atom in its cell at a distance r from its equilibrium position.

The thermodynamic properties can be determined from the translation partition function of the atoms of mass m, in their free volume, which is known from statistical mechanics (see, e.g., Gurney 1966):

$$Z = \left(\frac{2\pi mkT}{h^2}\right)^{3/2} V_F \qquad (3.77)$$

Vashchenko and Zubarev start from the Mie–Grüneisen definition of the thermodynamic gamma (2.60)

$$\gamma = \frac{VP_{th}}{E_{th}}$$

where P_{th} and E_{th} are the thermal pressure and thermal energy, respectively, which can be expressed in terms of the partition function Z:

$$P_{th} = kT \left(\frac{\partial \ln Z}{\partial V}\right)_T$$

$$E_{th} = kT^2 \left(\frac{\partial \ln Z}{\partial T}\right)_V$$

Hence, incidentally, another interesting formulation of the thermodynamic gamma

$$\gamma = \left(\frac{\partial \ln Z}{\partial \ln V}\right)_T \left(\frac{\partial \ln Z}{\partial \ln T}\right)_V^{-1} \qquad (3.78)$$

Inserting (3.77) into (3.78), we obtain

$$\gamma = \left(\frac{\partial \ln V_F}{\partial \ln V}\right)_T \left[\frac{3}{2} + \left(\frac{\partial \ln V_F}{\partial \ln T}\right)_V\right]^{-1} \qquad (3.79)$$

From (3.76), expanding $\chi(r)$ in powers of r for small displacements of the atoms, integrating and approximating the energy by the pair interaction potential, Vashchenko and Zubarev find

▶
$$\gamma_{FV} = -\frac{V}{2} \left[\frac{d^2(PV^{4/3})}{dV^2}\right] \left[\frac{d(PV^{4/3})}{dV}\right]^{-1} \qquad (3.80)$$

which is to be compared with Slater's gamma formulation (3.64).

Irvine and Stacey (1975) start from an analysis of the three-dimensional oscillations of atoms on a simple cubic lattice and calculate the thermal energy and thermal pressure in the classical limit to obtain the Mie–Grüneisen thermal gamma. The outline of their demonstration is given here.

Let us consider a pair of neighboring atoms vibrating independently in the three directions x, y, z. At equilibrium $(T = 0 \text{ K})$, the atomic separation is r_e in the x direction and the force constant of the bond is $F(r_e)$. At finite temperature, the atoms vibrate and their interatomic separation is $r = [(r_e + \Delta x)^2 + \Delta y^2 + \Delta z^2]^{1/2}$, where $\Delta x, \Delta y, \Delta z$ are the components of the elongation of the bond; as they are small compared to r_e, r^{-1} can be expanded to second order in $\Delta x, \Delta y, \Delta z$. The force constant can be expanded to second order in $(r - r_e)$, and its x-component $F_x = F(r)(r_e + \Delta x)r^{-1}$ is expressed to second order using the expansions of $F(r)$ and r^{-1}. Its time average $\langle F_x \rangle$ must be balanced by the force created by the external pressure P over the area r_e^2 so that the specific volume remains constant; P is then equal to the thermal pressure. In taking the time average of F_x, the time average $\langle \Delta x \rangle$ is taken equal to zero since the external pressure prevents thermal expansion and in the classical limit the potential energy of each bond is equal to $\frac{1}{2}kT$. The quadratic terms in the expansion of $\langle F_x \rangle$ can therefore be written as functions of r_e and of $F(r_e)$ and its derivatives with respect to r, taken at r_e: $F'(r_e)$, $F''(r_e)$. The thermal pressure appears in a Mie–Grüneisen-type equation, using the high-temperature classical Dulong and Petit value for the specific heat at constant volume: $C_V = 3k$ per atom. The Grüneisen parameter is calculated as:

$$\gamma = -\frac{1}{3}\left[\frac{1}{2}F''(r_e) + \frac{F'(r_e)}{r_e} - \frac{F(r_e)}{r_e^2}\right]\left[\frac{F'(r_e)}{r_e} + \frac{2F(r_e)}{r_e^2}\right]^{-1}$$

Since $P_0 = F(r_e)/r_e^2$, we can express r_e, $F(r_e)$, $F'(r_e)$, and $F''(r_e)$ in terms of P and the bulk modulus K and its derivative dK/dP.

The Grüneisen parameter can thus be written

$$\gamma_{VZ} = \left(\frac{1}{2}\frac{dK}{dP} - \frac{5}{6} + \frac{2}{9}\frac{P}{K}\right)\left(1 - \frac{4}{3}\frac{P}{K}\right)^{-1} \tag{3.81}$$

which is equivalent to (3.80) and can be easily derived from it by using the definition of the bulk modulus $K = -V\partial P/\partial V$. The calculation can be generalized to other lattices than the simple cubic one. The expression (3.81) still holds, as long as the interatomic forces are purely central.

Dugdale and MacDonald (1953) proposed another expression for the Grüneisen parameter correcting the alleged error made by Slater in neglecting to take into account the effect of finite strain under applied pressure:

$$\gamma_{DM} = \frac{1}{2}\left(\frac{dK}{dP} - 1 + \frac{2P}{9K}\right)\left(1 - \frac{2P}{3K}\right)^{-1} \tag{3.82}$$

or

$$\gamma_{DM} = -\frac{1}{3} - \frac{V}{2}\left[\frac{\partial^2(PV^{2/3})}{\partial V^2}\right]\left[\frac{\partial(PV^{2/3})}{\partial V}\right]^{-1} \tag{3.83}$$

However, Gilvarry (1956d) shows that use of the formal theory of finite strain also leads to Slater's gamma and that no physical basis exists for the modification proposed by Dugdale and MacDonald. Irvine and Stacey (1975) find that their derivation of the free-volume gamma leads to the Dugdale and MacDonald gamma if they assume one-dimensional atomic motions, thus displaying the weakness of the model.

N.B.   Incidentally, note that at $P = 0$, we have $\gamma_{DM} = \frac{1}{2}(\gamma_S + \gamma_{VZ})$.

The relations between various gammas are summarized in Figure 3.10.

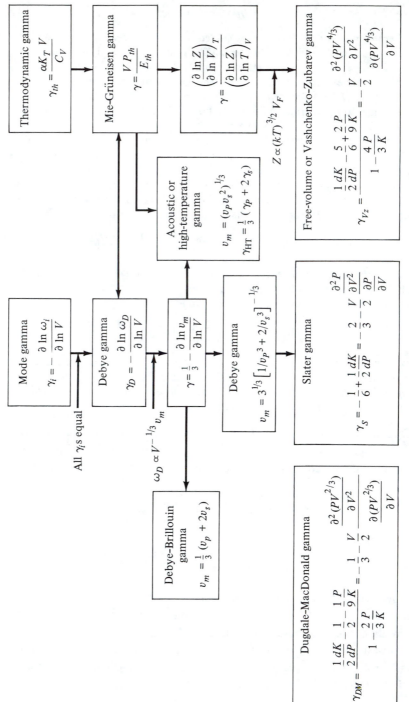

Figure 3.10. Relations among the various definitions of the Grüneisen parameter.

That Grüneisen's parameter varies with pressure clearly appears from the calculations, but the variation depends on how the sound velocities or the elastic moduli vary with pressure and the data do not always exist, or, if they exist, they are not always easy to extrapolate to high pressures. Depending on which equations of state (i.e., $K = f(P)$) are used, the results may differ widely (Irvine and Stacey 1975).

A useful empirical law relates the Grüneisen parameter to the density ρ (O. L. Anderson 1979):

$$\gamma \rho^q = \text{const.} \tag{3.84}$$

with values of q between 0.8 and 2.2 acceptable for the lower mantle.

Boehler and Ramakrishnan (1980) determined the pressure dependence of the thermodynamic gamma by measuring the temperature change associated with adiabatic compression up to 5 kbar, using the expression

$$\gamma_{th} = \frac{\alpha K_S}{\rho C_P} = \frac{K_S}{T}\left(\frac{\partial T}{\partial P}\right)_S$$

They found values of q between 0.6 (for Fe) and 1.7, with an average value of 1.3 for metals and alkali halides.

It is simpler, not necessarily more inaccurate in view of all the other sources of uncertainty, and often experimentally correct to assume that $q = 1$, hence

$$\gamma \rho = \gamma_0 \rho_0 = \text{const.} \tag{3.85}$$

The formula can be given some justification by considering the thermodynamic gamma $\gamma_{th} = \alpha K_T / \rho C_V$: at high temperature in the classical limit $C_V = 3nR$ and

$$\gamma \rho = \frac{\alpha K_T}{3nR} \tag{3.86}$$

Now, there is reasonable experimental evidence that αK_T remains constant at high temperatures for a number of crystals (Birch 1968; Brennan and Stacey 1979; Steinberg 1981).

The Grüneisen parameter can reasonably be considered as independent of temperature to first approximation (Irvine and Stacey 1975).

Table 3.3 gives values of the various gammas for a few important crystals and mantle minerals, from Quareni and Mulargia (1988) and D. L. Anderson (1988).

Two observations can be made:

i. For all materials, the values of gamma at ambient pressure roughly lie between 0.5 and 3 and are generally close to 2.

Table 3.3. *Grüneisen parameters of some typical crystals*

Crystal	γ_{th}	γ_a	γ_S	γ_{DM}	γ_{VZ}	γ_D	γ_{Br}	Source
MgO	1.52	1.41						A
	1.54		1.12	0.78	0.45	0.92	0.95	QM (a)
	1.54		1.89	1.55	1.22	1.40	1.48	QM (b)
Al_2O_3	1.13	1.15						A
	1.28		1.92	1.59	1.25	1.25	1.39	QM (b)
$MgAl_2O_4$	1.40	0.67						A
	1.40		2.18	1.84	1.51	0.44	0.85	QM (b)
Mg_2SiO_4	1.16	1.49						A
	1.18		2.39	2.05	1.72	1.26	1.51	QM (b)
$SrTiO_3$	1.63	1.96						A
Fe	1.65		3.14	2.80	2.47	2.26	2.48	QM (a)
	1.65		2.65	2.31	1.98	1.76	1.99	QM (b)

Notes: γ_{th} = thermodynamic gamma; γ_a = acoustic gamma; γ_S = Slater gamma; γ_{DM} = Dugdale–MacDonald gamma; γ_{VZ} = Vashchenko–Zubarev gamma; γ_D = Debye gamma; γ_{Br} = Debye–Brillouin gamma.
Sources: A: O. L. Anderson (1988). QM (a): Quareni and Mulargia (1988) (Hashin–Shtrickman). QM (b): Quareni and Mulargia (1988) (Voigt–Reuss–Hill).

ii. For the same material the values of gamma vary widely, depending of course on which gamma is calculated, but also on the origin of the experimental data used and even on the way the elastic data are averaged (Voigt–Reuss–Hill or Hashin–Shtrickman).

3.6 Harmonicity, anharmonicity, and quasi-harmonicity

3.6.1 Generalities

Let us come back to the schematic case of a lattice of atoms vibrating as harmonic oscillators about their equilibrium position. The restoring force $F = k(r - r_e)$ is proportional to the elongation $r - r_e$, and the symmetrical interatomic pair potential is parabolic (Section 3.2.1) (Fig. 3.11). This case corresponds to the linear infinitesimal elasticity, with elastic moduli independent of strain (or stress).

In reality, the equilibrium distance between atoms r_e is determined by the balance between a long-range attractive force and a short-range repulsive force. The interatomic potential $E(r)$ can be expressed as the sum of an attractive and a repulsive potential, for example, by Mie's formula:

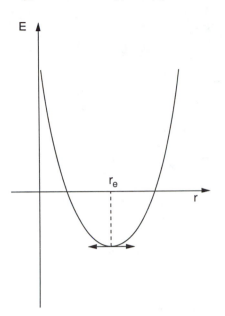

Figure 3.11. Harmonic curve of interatomic potential. The equilibrium inter-atomic distance r_e corresponds to the minimum of the parabolic curve.

$$E(r) = -\frac{a}{r^m} + \frac{b}{r^n} \tag{3.87}$$

with $n > m$, since the repulsive potential has a shorter range. A simple case is that of ionic crystals, for which $m = 1$ (Coulombic attraction) and $n \approx 9$ (Born potential). The curve of bond energy (Fig. 3.12) is therefore asymmetrical, steeper for smaller interatomic distances. This results in nonlinear anharmonic oscillations. As the interatomic spacing decreases with increasing applied pressure, the restoring force (i.e., the negative of the slope of the curve) increases more rapidly than in the harmonic case: Compression becomes more and more difficult, that is, the bulk modulus increases with pressure.

Of course, near its minimum the curve of bond energy can be approximated by its osculatory parabola (with the same curvature) and the linear approximation is valid.

3.6.2 Thermal expansion

Another important consequence of the asymmetry of the potential curve appears when the temperature is taken into consideration. At low temperatures, the quantized energy levels are near the bottom of the potential

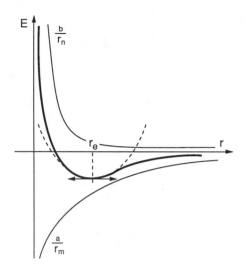

Figure 3.12. Anharmonic curve of interatomic potential (bold curve). The potential is the sum of the attractive Coulombic potential and a short-range repulsive potential (thin curves). The osculatory parabola at $r = r_e$ (dashed curve) corresponds to the harmonic approximation.

well and the vibrational amplitudes are small; the harmonic approximation holds. However, as temperature increases, the energy levels climb higher up in the well and the amplitudes increase, and the asymmetry of the well cannot be neglected any longer and causes the oscillations to become nonsinusoidal or, in other words, anharmonic.

The interatomic distance oscillates between r_1 and r_2, with $r_2 - r_e > r_e - r_1$, that is, the extension of the bond is greater than its compression (Fig. 3.13). In addition, the restoring force is smaller on the extension side, so that, on average, the bond spends a longer time in extension than in compression. The mean value of the bond length becomes longer than the equilibrium value at low temperature: $r_1 + r_2 > 2r$. This is the cause of thermal expansion. We have already seen (Section 2.3.3) that we can relate thermal expansion to thermal pressure by considering that heating a solid at constant volume increases the internal (negative) pressure, which, in turn, would cause the solid to expand if the constraint of constant volume were lifted:

$$\left(\frac{\partial P}{\partial T}\right)_V = \alpha K_T = \gamma_{th} \rho C_V$$

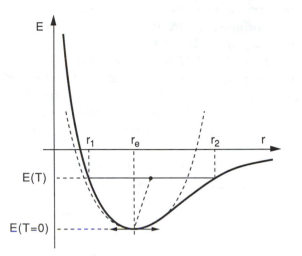

Figure 3.13. Thermal expansion. At $T=0$ K, the interatomic potential is minimum for the equilibrium distance r_e. At a finite temperature, the potential $E(T)$ is higher and due to the asymmetric shape of the anharmonic curve, the equilibrium distance is the average between r_1 and r_2, which is greater than r_e. For a symmetrical harmonic potential (dashed curve) there would be no thermal expansion.

The Grüneisen ratio is therefore a good parameter of anharmonicity and we have seen in Section 3.5 that it can be related to dK/dP, which is also linked to the bond asymmetry. It can also be expressed as a function of the exponents m and n of the bond potential in Mie's formula (3.87) (Zwikker 1954): By expanding the potential $E(r)$ in a series of powers of the vibrational elongations, the thermal energy can be expressed as a function of m and n and Grüneisen's parameter is found to be (see also Section 4.4):

$$\gamma_{th} = \frac{1}{2} + \frac{m+n}{6} \qquad (3.88)$$

Conversely, we can deal with the effect of temperature on the vibration modes and elastic moduli by considering that an increase in temperature causes an increase in volume by thermal expansion and calculating the effect of the increase in volume on the elastic properties through quantities such as $-d \ln K/d \ln V = dK/dP$. This is the so-called *quasi-harmonic approximation*, which takes into account the effect of temperature through the volume change due to thermal expansion only. In the quasi-harmonic approximation, the vibrations are considered to be harmonic about the new equilibrium positions of the atoms corresponding to the expanded state.

However, we must not overlook the fact that there are anharmonic effects due to temperature that are not accounted for by the quasi-harmonic approximation (e.g., changes in the electronic configuration of the bonds). D. L. Anderson (1987a, 1988) distinguishes between the *intrinsic* derivatives of the elastic moduli, expressing the effect of temperature or pressure at constant volume, and the *extrinsic,* or volume-dependent, derivatives. They are conveniently expressed as dimensionless logarithmic derivatives with respect to specific mass ρ (Dimensionless Logarithmic Anharmonic or DLA parameters).

An often quoted DLA parameter (D. L. Anderson 1987a) is the *Grün-eisen–Anderson* (O. Anderson) *parameter* δ_T or δ_S:

$$\delta_T = \left(\frac{\partial \ln K_T}{\partial \ln \rho}\right)_P = -\frac{1}{\alpha}\left(\frac{\partial \ln K_T}{\partial T}\right)_P$$

$$\delta_S = \left(\frac{\partial \ln K_S}{\partial \ln \rho}\right)_P = -\frac{1}{\alpha}\left(\frac{\partial \ln K_S}{\partial T}\right)_P$$

(3.89)

It is possible to introduce an intrinsic anharmonic parameter a_i for every vibration mode, which can be experimentally determined by measuring separately the variation with temperature and pressure of the mode frequencies by infrared or Raman spectroscopy (Gillet, Guyot, and Malezieux 1989). There are therefore two mode Grüneisen parameters:

$$\gamma_{i,T} = -\left(\frac{\partial \ln \omega_i}{\partial \ln V}\right)_T \quad \text{and} \quad \gamma_{i,P} = -\left(\frac{\partial \ln \omega_i}{\partial \ln V}\right)_P$$

We can write

$$d \ln \omega_i = a_i dT - \gamma_{i,T} d \ln V = a_i \frac{\partial T}{\partial P} dP + \left(a_i \frac{\partial T}{\partial \ln V} - \gamma_{i,T}\right) d \ln V$$

Hence

$$\left(\frac{d \ln \omega_i}{d \ln V}\right)_P = \gamma_{i,P} = a_i \frac{\partial T}{\partial \ln V} - \gamma_{i,T}$$

or

$$a_i = \left(\frac{\partial \ln \omega_i}{\partial T}\right)_V = \alpha(\gamma_{i,T} - \gamma_{i,P})$$

(3.90)

Chopelas and Boehler (1989) determined the pressure dependence of the thermal expansion coefficient of MgO and forsterite by putting experimentally determined values of thermodynamic parameters in the equation

$$\alpha = \frac{K_S C_V}{K_T V T}\left(\frac{\partial T}{\partial P}\right)_S$$

They found values of $(\partial \ln \alpha / \partial \ln \rho)_T$ equal to -6.5 and -6.1 for MgO and forsterite, respectively. A systematics of $(\partial \ln \alpha / \partial \ln \rho)_T$ for many minerals yields an average value of 5.5 ± 0.5.

4

Equations of state

4.1 Generalities

The thermodynamic state of a system is usually defined by pressure P, temperature T, and specific volume V (or specific mass ρ), linked by the *equation of state* (EOS) (see, e.g., Eliezer, Ghatak, and Hora 1986). The best known EOS is the one for ideal gases:

$$PV = RT$$

We are concerned here with the various possible EOS for solids at high pressures and high temperatures (see Zharkov and Kalinin 1971), for, in order to build compositional (mineralogical) Earth models, it is indispensable to have an EOS for the regions of the interior of the Earth, as well as for the candidate minerals: Inversion of seismic travel times yields seismic velocities at all depths (pressures), and to compare the velocities, or the inferred values of densities or elastic moduli, with those of mineral assemblages, one needs either to "compress" the assemblages to high pressures or "decompress" the Earth material to ambient pressure. Hence the need for an EOS.

Since in the case of solids the effect of temperature is much less than for gases, it is often introduced only as a thermal expansion correction to the *isothermal EOS*, $V(P)$ or $\rho(P)$, which is the one usually experimentally determined for minerals at high pressures (Stacey, Brennan, and Irvine 1981). We will therefore start by considering isothermal equations of state (see reviews by Boschi and Caputo 1969; Stacey et al. 1981).

The simplest isothermal EOS one can think of, for solids, is given by the definition of bulk modulus K:

$$K = -\frac{dP}{d \ln V} = \frac{dP}{d \ln \rho} \qquad (4.1)$$

If we consider infinitesimal strains resulting from the application of hydrostatic pressure to an initially unstressed solid, we are in the case of linear elasticity with constant bulk modulus, and the EOS is obtained by integrating (4.1) with $K = K_0$:

$$V = V_0 \exp\left(\frac{-P}{K_0}\right) \tag{4.2}$$

This simple EOS is obviously not correct for high pressures since it does not take into account the fact that it becomes more and more difficult to compress the solid, that is, that the bulk modulus increases with pressure. The seismic waves in the Earth, as well as the ultrasonic waves in laboratory experiments, propagate in media that have undergone a finite strain under high pressure, and even though the strain corresponding to the propagation of the waves can be considered infinitesimal and treated in the linear approximation, we must not neglect the fact that the large finite strain has modified the elastic moduli of the medium.

4.2 Murnaghan integrated linear equation of state

Let us consider infinitesimal strains resulting from the application of hydrostatic pressure P to a solid initially compressed to a finite strain by a pressure P_0.

Murnaghan (1967) demonstrated that the effective local bulk modulus is (see also Sec. 4.3.2 N.B.):

$$K = \tfrac{1}{3}(3\lambda + 2\mu + P_0) \tag{4.3}$$

In the case of isotropic solids, this is equivalent to having 3 elastic constants, λ, μ, and P_0. λ and μ depend on the initial pressure P_0 and are equal to λ_0 and μ_0 (Lamé constants of the solid in the natural state) for $P_0 = 0$. If we make the assumption that λ and μ are linear functions of P_0, the local bulk modulus is also a linear function of P_0:

$$K = \tfrac{1}{3}(3\lambda_0 + 2\mu_0) + kP_0 = K_0 + kP_0 \tag{4.4}$$

Using (4.1), we have:

$$d \ln \rho = \frac{dP}{K_0 + kP_0}$$

Since P_0 is arbitrary, it can be replaced by P and we obtain by integration

$$P = \frac{K_0}{k}\left[\left(\frac{\rho}{\rho_0}\right)^k - 1\right] \tag{4.5}$$

If we made the drastic assumption that λ and μ do not depend on P_0, we would have $k = \tfrac{1}{3}$. If, now, we assume that the dependence of K on the pre-applied pressure is the same as that of K_0 on pressure about $P = 0$, that is, if we replace k with $K_0' = (dK/dP)_{P=0}$, we can write (4.5) as:

▶
$$P = \frac{K_0}{K_0'} \left[\left(\frac{\rho}{\rho_0} \right)^{K_0'} - 1 \right] \tag{4.6}$$

which can also be written

$$\rho = \rho_0 \left(1 + \frac{K_0'}{K_0} P \right)^{1/K_0'} \tag{4.7}$$

This is *Murnaghan integrated linear EOS* (Murnaghan 1967), usually simply called *Murnaghan EOS*.

An equation formally identical to (4.6) could be derived by expanding K to first order in P about $P = 0$,

$$K \approx K_0 + K_0' P \tag{4.8}$$

inserting (4.1) into (4.8), and integrating, but this apparently simpler procedure is not physically rigorous and hides the important role of the finite strain.

4.3 Birch–Murnaghan equation of state

4.3.1 Finite strain

The EOS most currently used, especially in the treatment of experimental compression data of minerals, is the Eulerian finite strain Birch–Murnaghan EOS. In view of its importance, we will devote some space to presenting it here, first introducing the necessary background, starting with the expression for finite strain.

Let us consider a solid in the coordinate axes $Ox_1 x_2 x_3$ (Fig. 4.1), and let ds be the distance between two neighboring points P of coordinates x_i ($i = 1, 2, 3$) and Q of coordinates $x_i + dx_j$. We have

$$ds^2 = \sum_i (dx_i)^2 \tag{4.9}$$

Let points P and Q be displaced to P' and Q' by a *displacement* vector $\mathbf{u}(x_i)$, which is a function of the coordinates of the points.

$$P(x_i) \rightarrow P'(x_i + u_i)$$

$$Q(x_i + dx_i) \rightarrow Q'(x_i + dx_i + u_i + du_i)$$

Since $\mathbf{u}$ is not constant for all points, which would correspond to a rigid-body translation, the distance dS between P' and Q' is different from ds: The solid undergoes a *strain*.

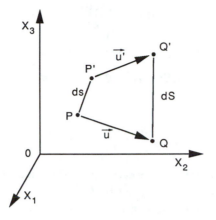

Figure 4.1. The distance *ds* between two neighboring points *P, P'* becomes *dS* after straining, due to the nonuniformity of the displacement field **u**.

To define the notion of strain and perform all the necessary calculations, we can use either of two schemes.

The *Lagrangian* scheme, in which one expresses the coordinates X_i of a point in the strained state as a function of its coordinates x_i in the initial, unstrained state:

$$X_i = x_i + u_i \tag{4.10}$$

The *Eulerian* scheme, in which one expresses the coordinates x_i of a point in the initial unstrained state as a function of its coordinates X_i in the strained state:

$$x_i = X_i - u_i \tag{4.11}$$

The two schemes yield definitions of strain that are equivalent for infinitesimal strain but not for finite strain. The Eulerian scheme is more physically meaningful for finite strain conditions, since every quantity is expressed in terms of the coordinates of the points in the strained solid, which are the experimentally accessible ones. As a consequence, the Eulerian scheme is commonly used to determine experimental equations of state and we will adopt it in what follows.

We have

$$dS^2 - ds^2 = \sum_i (dX_i)^2 - \sum_i (dx_i)^2 = 2 \sum_i (dX_i)(du_i) - \sum_i (du_i)^2 \tag{4.12}$$

If the displacements **u** are continuous and differentiable functions of x_i, we have

$$du_i = \sum_k \frac{\partial u_i}{\partial X_k} \, dX_k \tag{4.13}$$

The Eulerian finite strain tensor ϵ_{ij} is defined by

$$dS^2 - ds^2 = 2 \sum_{ij} \epsilon_{ij} \, dX_i \, dX_j \tag{4.14}$$

with

▶ $$\epsilon_{ij} = \frac{1}{2} \left(\frac{\partial u_i}{\partial X_j} + \frac{\partial u_j}{\partial X_i} \right) - \frac{1}{2} \sum_k \frac{\partial u_k}{\partial X_i} \frac{\partial u_k}{\partial X_j} \tag{4.15}$$

For infinitesimal strains, we can neglect the quadratic term and we obtain the definition of strain given by (2.1).

The Eulerian strain is sometimes written

$$\epsilon_{ij} = \frac{1}{2} \left(\delta_{ij} - \sum_k \frac{\partial x_k}{\partial X_i} \frac{\partial x_k}{\partial X_j} \right) \tag{4.16}$$

where $\delta_{ij} = 1$ if $i = j$ and $\delta_{ij} = 0$ if $i \neq j$.

N.B.   It is easy to show that (4.16) is equivalent to (4.15) by carrying $x_k = X_k - u_k$ into (4.16). We have then

$$2\epsilon_{ij} = \delta_{ij} - \sum_k \left(\frac{\partial X_k}{\partial X_i} \frac{\partial X_k}{\partial X_j} + \frac{\partial X_k}{\partial X_i} \frac{\partial u_k}{\partial X_j} \right.$$

$$\left. + \frac{\partial X_k}{\partial X_j} \frac{\partial u_k}{\partial X_i} - \frac{\partial u_k}{\partial X_i} \frac{\partial u_k}{\partial X_j} \right)$$

$$2\epsilon_{ij} = \delta_{ij} - \delta_{ij} + \delta_{ki} \frac{\partial u_k}{\partial X_j} + \delta_{kj} \frac{\partial u_k}{\partial X_i} - \sum_k \frac{\partial u_k}{\partial X_i} \frac{\partial u_k}{\partial X_j}$$

which can be immediately reduced to (4.15).

N.B.   The Lagrangian finite strain is

$$e_{ij} = \frac{1}{2} \left(\sum_k \frac{\partial X_k}{\partial x_i} \frac{\partial X_k}{\partial x_j} - \delta_{ij} \right) \tag{4.17}$$

Let us now focus our attention on the isotropic compressional strain caused by the application of hydrostatic pressure and let us write:

$$\frac{\partial u_1}{\partial X_1} = \frac{\partial u_2}{\partial X_2} = \frac{\partial u_3}{\partial X_3} = \frac{\theta}{3} \tag{4.18}$$

where $\theta = \sum_i \partial u_i / \partial X_i = \Delta V / V_0$ is the trace of the infinitesimal strain tensor. From (4.15) and (4.18), we have

$$\epsilon_{ij} = \epsilon \delta_{ij}$$

with

$$\epsilon = \frac{\theta}{3} - \frac{1}{2}\frac{\theta^2}{9} \tag{4.19}$$

An elementary cube of volume $V = (dX_1)^3$ in the strained state has a volume equal to $V_0 = [dX_1(1 - \partial u_1/\partial X_1)]^3$ in the unstrained state, hence

$$\frac{V_0}{V} = \frac{\rho}{\rho_0} = \left(1 - \frac{\partial u_1}{\partial X_1}\right)^3 = \left(1 - \frac{\theta}{3}\right)^3$$

Using the simple trick of writing

$$\left(1 - \frac{\theta}{3}\right)^3 = \left[\left(1 - \frac{\theta}{3}\right)^2\right]^{3/2}$$

we see that

$$\frac{\rho}{\rho_0} = \left(1 - \frac{2\theta}{3} + \frac{\theta^2}{9}\right)^{3/2} = \left[1 - 2\left(\frac{\theta}{3} - \frac{\theta^2}{18}\right)\right]^{3/2} = (1 - 2\epsilon)^{3/2} \tag{4.20}$$

As ϵ (dilatation) is negative for positive pressures, we introduce the "compression" $f = -\epsilon$.

$$\frac{\rho}{\rho_0} = \frac{V_0}{V} = (1 + 2f)^{3/2} \tag{4.21}$$

We see that for infinitesimal strains, we obtain the well-known result:

$$\frac{\rho}{\rho_0} = \frac{V_0}{V} \approx 1 - \theta \tag{4.22}$$

The isothermal bulk modulus at $P = 0$ is

$$K_{0T} = -\lim_{P \to 0}\left(\frac{PV}{\Delta V}\right)_T = -\lim_{P \to 0}\left(\frac{P}{\theta}\right)_T \tag{4.23}$$

and, since $P = -(\partial F/\partial V)_T$:

$$K_{0T} = \lim_{P \to 0}\left(\frac{1}{\theta}\frac{\partial F}{\partial V}\right)_T \tag{4.24}$$

Now, to first order, we have

$$\frac{V}{V_0} = (1 - 2f)^{-3/2} \approx 1 - 3f \tag{4.25}$$

and

$$dV \approx -3V_0 df$$

Since f tends toward zero with P, we can write:

$$9K_{0T}V_0 = \lim_{P \to 0} \left(\frac{1}{f} \frac{\partial F}{\partial f} \right)_T \qquad (4.26)$$

Let us now expand the free energy F in powers of f. If we take the energy of the unstrained state equal to zero and remember that the elastic strain energy is quadratic for infinitesimal strains, we have

$$F = a(T)f^2 + b(T)f^3 + c(T)f^4 + \cdots \qquad (4.27)$$

4.3.2 Second-order Birch–Murnaghan equation of state

We will follow Bullen (1975) for the demonstration of the Birch–Murnaghan EOS:

Let us consider only the expansion of F to second order:

$$F \approx af^2 \qquad (4.28)$$

From (4.26) and (4.27), we obtain

$$a = \tfrac{9}{2} K_{0T} V_0 \qquad (4.29)$$

The pressure is

$$P = -\left(\frac{\partial F}{\partial V} \right)_T = -\left(\frac{dF}{df} \right)_T \frac{df}{dV} \qquad (4.30)$$

Differentiating (4.21), we get

$$\frac{df}{dV} = -\frac{1}{3V_0}(1 + 2f)^{5/2} \qquad (4.31)$$

Hence

$$P = 3K_{0T}f(1 + 2f)^{5/2} \qquad (4.32)$$

Using (4.21), we can express f in terms of ρ/ρ_0,

$$f = \frac{1}{2}\left[\left(\frac{\rho}{\rho_0} \right)^{2/3} - 1 \right] \qquad (4.33)$$

and carrying it into (4.32), we obtain the second-order Birch–Murnaghan equation of state (Birch 1938, 1947):

$$\blacktriangleright \qquad P = \frac{3K_{0T}}{2}\left[\left(\frac{\rho}{\rho_0} \right)^{7/3} - \left(\frac{\rho}{\rho_0} \right)^{5/3} \right] \qquad (4.34)$$

The bulk modulus can be directly calculated from (4.34)

$$K = \frac{\rho}{\rho_0} dP \left(d \frac{\rho}{\rho_0} \right)^{-1} \tag{4.35}$$

$$K = \frac{K_{0T}}{2} \left[7 \left(\frac{\rho}{\rho_0} \right)^{7/3} - 5 \left(\frac{\rho}{\rho_0} \right)^{5/3} \right] \tag{4.36}$$

and using (4.21) again

$$K = K_{0T}(1+7f)(1+2f)^{5/2} \tag{4.37}$$

The seismic parameter Φ is, then, a simple polynomial in f:

$$\Phi = \frac{K}{\rho} = \frac{K_0}{\rho_0}(1+2f)(1+7f)$$

or

$$\Phi = \Phi_0(1+9f+14f^2) \tag{4.38}$$

The pressure derivative of the bulk modulus is calculated from (4.37) and (4.31):

$$\left(\frac{dK}{dP} \right)_{P=0} = K_0' = \frac{dK}{df} \frac{df}{dP}$$

$$K_0' = (12+49f)(3+21f)^{-1} \tag{4.39}$$

It is interesting to note that purely from finite strain theory, we obtain the numerical value $K_0' = 4$, for the infinitesimal case $f \to 0$. The experimental values of K_0' for many close-packed minerals are usually close to 4 (Table 4.1).

Note also that $K_0' = 4$ corresponds to reasonable values of the Grüneisen parameters

$$\gamma_S = 1.8 \qquad \gamma_{VZ} = 1.2$$

Introducing the value $K_0' = 4$ in Murnaghan's equation of state (4.6), we can compare the pressure P' it predicts with the pressure P predicted by the Birch–Murnaghan EOS (4.34) for the same values of ρ/ρ_0. We see (Fig. 4.2) that P' is slightly greater than P, but for values of $\rho/\rho_0 < 1.25$ given by the seismological PREM model for the lower mantle, the relative difference is about 3 percent.

N.B. Let us consider the results we obtain if, expanding, as we have done in equation (4.29), the strain energy to second order in the finite strain f, we calculate P, K, and K_0' in the limit $f \to 0$. We see from (4.32) that $P = 3fK_{0T}$ when $f \to 0$, and we have from (4.22) that $\rho/\rho_0 = 1+3f$. Hence

Table 4.1. *Pressure derivative at room temperature of the bulk moduli of close-packed crystals*

Crystal	$\left(\dfrac{dK}{dP}\right)_{P=0}$	Reference
MgO	4.1	Jackson and Niesler (1982)
CaO	4.2	Richet et al. (1988)
Mg_2GeO_4 spinel	4.1	Rigden et al. (1988)
Mg_2SiO_4 forsterite	4.8	Graham and Barsch (1969)
	5.4	Kumazawa and Anderson (1969)
	4.0	Will et al. (1986)
Fe_2SiO_4 fayalite	5.2	Graham et al. (1988)
$(Mg_{0.8}Fe_{0.2})SiO_3$ majorite	4.4	Jeanloz (1981)
$MgSiO_3$ perovskite	3.9	Knittle and Jeanloz (1987)

$$P = K_{0T}\left(\frac{\rho}{\rho_0} - 1\right)$$

$$K = \rho\frac{\partial P}{\partial \rho} = K_{0T}\frac{\rho}{\rho_0} = K_{0T} + P$$

This physically corresponds to a situation where, having applied pressure P and having compressed the solid to a finite compression f, we then increase the strain f by infinitesimal increments. The effective bulk modulus is then increased by P (see eq. (4.3)).

4.3.3 Third-order Birch–Murnaghan equation of state

If we expand F to third order in f,

$$F = a(T)f^2 + b(T)f^3 \tag{4.40}$$

we have

$$P = 3K_{0T}f(1+2f)^{5/2}\left(1 + \frac{3bf}{2a}\right) \tag{4.41}$$

After a tedious calculation, we obtain the third-order Birch–Murnaghan EOS

$$\blacktriangleright \quad P = \frac{3K_{0T}}{2}\left[\left(\frac{\rho}{\rho_0}\right)^{7/3} - \left(\frac{\rho}{\rho_0}\right)^{5/3}\right]\left\{1 + \frac{3}{4}(K_0' - 4)\left[\left(\frac{\rho}{\rho_0}\right)^{2/3} - 1\right]\right\} \tag{4.42}$$

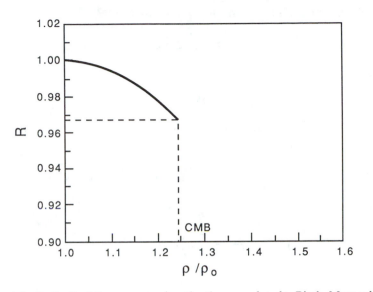

Figure 4.2. Ratio R of the pressure given by the second-order Birch–Murnaghan equation of state to the pressure given by Murnaghan's equation (with $K_0' = 4$) as a function of ρ/ρ_0 (ratio of the density to the density at zero pressure) for the material of the lower mantle. At the core–mantle boundary $R = 0.968$.

which is usually written

$$P = \frac{3K_{0T}}{2}\left[\left(\frac{\rho}{\rho_0}\right)^{7/3} - \left(\frac{\rho}{\rho_0}\right)^{5/3}\right]\left\{1 - \xi\left[\left(\frac{\rho}{\rho_0}\right)^{2/3} - 1\right]\right\}$$

Note that if $K_0' = 4$, we obtain the second-order equation (4.43).

Static equations of state of minerals are currently obtained up to pressures of several hundred kilobars (e.g., Will et al. 1986) and even above the megabar (Knittle and Jeanloz, 1987). The method consists in measuring the lattice parameters by X-ray diffraction in situ in a diamond-anvil cell at various pressures and fitting the data by a third-order Birch–Murnaghan equation. The errors involved in such parameter fitting have been analyzed by Bell, Mao, and Xu (1987).

4.4 Equations of state derived from interatomic potentials

Equations of state of the Birch–Murnaghan type, with various values of the exponents of (ρ/ρ_0), can be justified by the choice of appropriate interatomic attractive and repulsive potentials (see Section 3.6) (Gilvarry 1957a; Stacey et al. 1981).

Let us, for example, consider Mie's equation (3.87)

$$E = -\frac{a}{r^m} + \frac{b}{r^n} \tag{4.43}$$

which we can write as

$$E = -aV^{-m/3} + bV^{-n/3} \tag{4.44}$$

For the isothermal EOS at room temperature, the thermal energy can be neglected and the pressure is

$$P = -\left(\frac{\partial E}{\partial V}\right)_T = -\frac{ma}{3}V^{-(m/3+1)} + \frac{nb}{3}V^{-(n/3+1)} \tag{4.45}$$

and

$$K = -V\left(\frac{\partial P}{\partial V}\right)_T = -\frac{ma}{3}\left(\frac{m}{3}+1\right)V^{-(m/3+1)} + \frac{nb}{3}\left(\frac{n}{3}+1\right)V^{-(n/3+1)} \tag{4.46}$$

Setting $P = 0$ in (4.45) and $K = K_0$ in (4.46), we obtain

$$a = \frac{9K_0}{m(n-m)}V_0^{(m/3+1)}$$

$$b = \frac{9K_0}{n(n-m)}V_0^{(n/3+1)}$$

Hence

$$P = \frac{3K_0}{n-m}\left[\left(\frac{V_0}{V}\right)^{(n/3+1)} - \left(\frac{V_0}{V}\right)^{(m/3+1)}\right] \tag{4.47}$$

or

$$P = \frac{3K_0}{n-m}\left[\left(\frac{\rho}{\rho_0}\right)^{(n/3+1)} - \left(\frac{\rho}{\rho_0}\right)^{(m/3+1)}\right] \tag{4.48}$$

$$K = \frac{K_0}{n-m}\left[(n+3)\left(\frac{V_0}{V}\right)^{(n/3+1)} - (m+3)\left(\frac{V_0}{V}\right)^{(m/3+1)}\right] \tag{4.49}$$

$$K_0' = \left(\frac{dK}{dP}\right)_{P=0} = \frac{1}{3}(m+n+6) \tag{4.50}$$

With (4.50) and using Slater's relation (3.65),

$$\gamma_S = -\frac{1}{6} + \frac{1}{2}\frac{dK}{dP}$$

we find an expression for Slater's Grüneisen parameter:

$$\gamma_S = \frac{1}{6}(m+n+5) \tag{4.51}$$

We see that the second-order Birch–Murnaghan equation is obtained for $m = 2$ and $n = 4$ and that, as expected, we have $K_0' = 4$; Slater's gamma is then equal to 1.83.

Incidentally, if we compare equation (4.50) to equation (3.88) for the thermodynamic Grüneisen ratio calculated with the same approach, we see that

$$\gamma = \frac{1}{2} \frac{dK}{dP} - \frac{1}{2} \qquad (4.52)$$

which is equal to the Dugdale–MacDonald gamma for $P = 0$.

Other equations of state, derived from various more or less empirical interatomic potentials, are given in Stacey et al. (1981).

4.5 Birch's law and velocity–density systematics

4.5.1 Generalities

Having measured the compressional wave velocities v_P of some 250 specimens of rocks by ultrasonic methods, up to 10 kbars, Birch (1961a) found that at pressures above a few kilobars (when most cracks are closed), the principal factors determining velocity were the specific mass ρ and the mean atomic mass $\bar{M}$ (usually, although improperly, called "density" and mean atomic "weight," respectively). The mean atomic mass is equal to the sum of the atomic masses of all atoms in a formula unit, divided by the number of atoms. The mean atomic mass of Mg_2SiO_4, for instance, is equal to $M = (2 \times 24.3 + 28.3 + 4 \times 16)/7 = 20.13$ g. It turns out that most close-packed mantle oxides and silicates have a value of $\bar{M}$ close to 20 (e.g., 20.12 for $MgSiO_3$, 20.15 for MgO). In most cases, an increase in $\bar{M}$ above these values corresponds to the replacement of magnesium by the heavier iron. One can consider that for the mantle $\bar{M} = 21.1$, and for the core $\bar{M} = 49.3$.

Birch (1961a) was able to fit the data corresponding to the same $\bar{M}$ by a linear law (Fig. 4.3) and, for values corresponding to the mantle ($20 \le \bar{M} \le 22$), he found:

$$v_P = -1.87 + 3.05\rho \qquad (4.53)$$

with v_P in km/s and ρ in g/cm^3.

More generally, Birch (1961b) noted, "The experimental data suggest that, regardless of crystal structure, the representative points on the velocity–density diagram will move along a line of constant mean atomic weight," and that, in general,

$$v_P = a(\bar{M}) + b\rho \qquad (4.54)$$

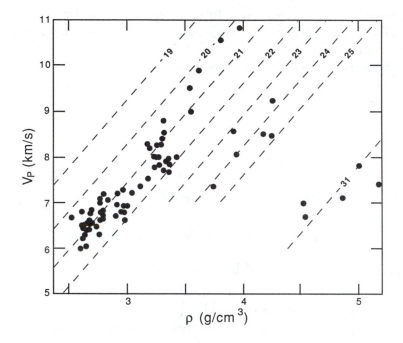

Figure 4.3. Velocity of the *P* waves at 10 kbars vs. specific mass for silicates and oxides. The dashed lines show the trend for a constant mean atomic mass (after Birch 1961a).

Also, plots of isostructural compounds lie on lines roughly perpendicular to the iso-$\bar{M}$ lines.

Birch wrote further: "*As a provisional hypothesis, it will be postulated that the velocity density relations, as expressed above, hold for all changes of density in the Earth's mantle, however produced* [italics added]." In other words, the changes in density produced by compression or by replacing a mineral by an "analogue" compound of same $\bar{M}$ but of different chemical composition, cause the same change in elastic wave velocity. Thirty years later, the "provisional hypothesis" still holds and relation (4.54), known as *Birch's law,* has given rise to the vast field of velocity-density systematics on analogue compounds.

Analogue compounds of minerals relevant to geophysics (mostly silicates) are usually isostructural crystals (more on the importance of structure in what follows) with different cations: in the perovskite structure, for instance, silicon can be replaced by germanium, titanium, or even aluminum, and magnesium and iron can be replaced by other alkaline earth or transition elements (Liebermann, Jones, and Ringwood 1977).

The replacement of divalent oxygen by the univalent fluoride anion gives rise to another class of analogues: fluorides, with weaker ionic bonds, are indeed rather good models of oxides (e.g., BeF_2 for SiO_2) and have been used to infer their elastic or thermal properties (Jones and Liebermann 1974; Jackson 1977).

The velocity–density systematics are currently used for three main purposes (Liebermann 1973):

Determining the density profiles from measured seismic velocities in regions of the mantle assumed to be chemically homogeneous.

Assessing the chemical homogeneity of regions of the mantle and relating the inhomogeneity to changes in chemical composition.

Relating the velocity jump at seismic discontinuities to the density increase known to occur for certain phase transitions.

The contribution of this approach to geophysics has been immense, but specific conclusions are usually not unique and must sometimes be taken with a grain of salt. Indeed, Schreiber and Anderson (1970) found that Birch's curves for various rocks such as diabase, gabbro, and eclogite fitted the data on various cheeses extremely well and that it was not possible to reject on acoustic grounds the hypothesis that the Moon is made of green cheese.

As the elastic wave velocities depend on the elastic constants and density, Birch's law is indeed an equation of state, if the changes in density are assumed to be produced by hydrostatic pressure. Furthermore, if one justifiably assumes that the velocities are insensitive to frequency to first order, that is, that the velocity–density relationship established in the laboratory at ultrasonic frequencies (≈ 1 MHz) is valid for seismic frequencies (≤ 10 Hz), then we have a *seismic equation of state* (D. L. Anderson 1967).

In what follows, we will try to extract the most important results from a considerable amount of published literature and, for details, we will refer the reader to the original papers and review articles (e.g., Wang 1978; Liebermann 1973, 1982; O. L. Anderson 1988). We will deal only with bulk-velocity–density systematics, since, despite some attempts (Davies 1976), the systematics on shear velocity are extremely unreliable, due to the great influence of the structure-sensitive noncentral bonding forces in shear.

4.5.2 Bulk-velocity–density systematics

Using published values of the bulk (or hydrodynamical) velocity $v_\Phi = \Phi^{1/2} = (K/\rho)^{1/2}$, obtained by shock-wave experiments or isothermal compression on metals, Birch (1961b, 1963) found that the bulk-velocity–density

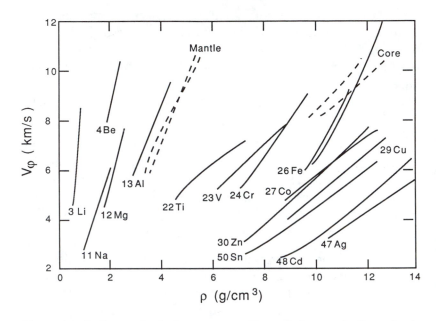

Figure 4.4. Bulk sound velocity versus specific mass for metals, from shock data (solid curves). The corresponding atomic masses are indicated. The curves (dashed) for the mantle and core materials, from seismic data, are also shown (after Birch 1963).

plots occurred in a sequence corresponding to the sequence of the atomic numbers of the metals (Fig. 4.4). Superimposing on the plots of the metals the plots corresponding to the mantle and to the core obtained from seismic data, Birch was led to the conclusion that the mantle was composed of light elements and that the core was essentially made of iron. McQueen, Fritz, and Marsh (1964), using shock-wave data, obtained a bulk-velocity–density Birch's diagram for oxides and silicates (Fig. 4.5).

Wang (1968a) used an experimental shock-wave equation of state to show that the bulk-velocity–density plot for MgO was indeed linear up to 1.26 Mbars, in agreement with Birch's postulate. He established a systematics similar to that of McQueen et al. (1964) with an equation of the form

$$v_\Phi = a(\bar{M}) + b\rho \tag{4.55}$$

For the values $20 \leq \bar{M} \leq 22$ corresponding to mantle minerals, Wang proposed that

$$v_\Phi = -1.75 + 2.36\rho \tag{4.56}$$

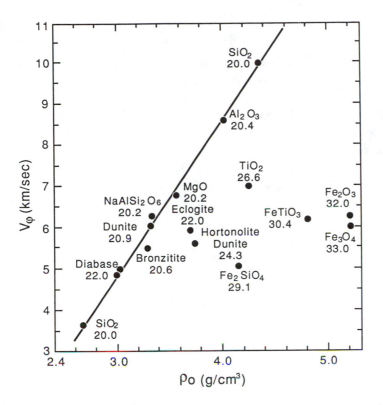

Figure 4.5. Bulk sound velocity versus specific mass for minerals and rocks of the mantle and crust. The mean atomic mass is indicated. The solid line is the locus of the points corresponding to material of mean atomic mass 20.1 (after McQueen et al. 1964).

Wang (1970) compared the plots of Φ/Φ_0 versus ρ/ρ_0 corresponding to the empirical relationship (4.56) with the plot derived from the Birch–Murnaghan expression of the seismic parameter

$$\frac{\Phi}{\Phi_0} = 7\left(\frac{\rho}{\rho_0}\right)^{4/3} - 5\left(\frac{\rho}{\rho_0}\right)^{2/3} \tag{4.57}$$

He found that the values of ρ/ρ_0 at given Φ/Φ_0, predicted by the linear relation, with coefficients consistent with experimental data were only 1 percent higher than those predicted by the Birch–Murnaghan EOS.

Shankland (1972, 1977) and Chung (1972) showed that Birch's law is consistent with Debye's theory: The sound velocity is $v = \omega/k$ and we have, from (3.28) and (3.56),

$$\frac{d \ln v}{d \ln \rho} = \gamma - \frac{1}{3} \qquad (4.58)$$

Since the values of Grüneisen's ratio for most materials are in the range of 1 to 2, the slopes of the constant mean atomic mass lines in Birch's diagram are roughly identical as a consequence of the near constancy of γ for the "isochemical" compounds. Integration of (4.58) leads to a power law (see 4.60) and Chung (1972) showed that Birch's law is a linearization of the power law over a certain range of densities corresponding to those of most rocks and minerals.

Chung (1972) and Shankland (1972), using a reasoning based on interatomic potentials for ionic crystals, also proposed that the isostructural variation of the bulk velocity with $\bar{M}$ is given by

$$\left(\frac{\partial \ln v_\Phi}{\partial \ln \bar{M}} \right)_{\text{Struct.}} \cong -\frac{1}{2}$$

or

$$v_\Phi \bar{M}^{1/2} = \text{const.} \qquad (4.59)$$

This relation can be justified by considering the acoustic velocity

$$v = \frac{\omega}{k} \propto \omega R_0$$

where R_0 is the interatomic distance at equilibrium, and by remembering that for harmonic atomic oscillations, the frequency ω is given by

$$\omega \propto \left(\frac{E''}{m} \right)^{1/2}$$

where E'' is the second derivative of the cohesive energy with respect to interatomic distance and m is the mass (see Section 3.2.1). For interatomic potentials appropriate for ionic crystals, such as (3.15), the cohesive energy is given by (3.18) and we have

$$E'' = \frac{N\alpha q^2(n-1)}{R_0^3}$$

Hence

$$v \propto \left[\frac{N\alpha q^2(n-1)}{mR_0} \right]^{1/2} \qquad (4.60)$$

Since the quantity $\alpha q^2(n-1)/R_0$ is practically constant for a given structure and $\bar{M}$ is proportional to m, it follows that the logarithmic derivative

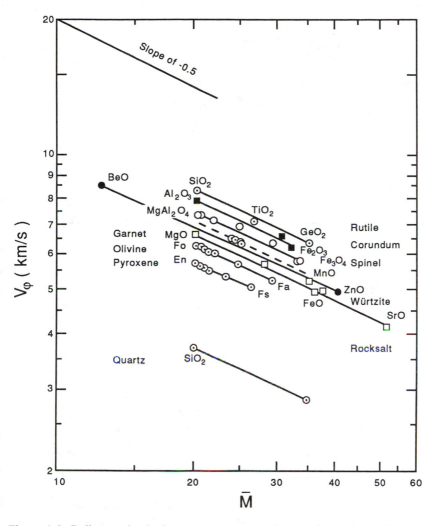

Figure 4.6. Bulk sound velocity versus mean atomic mass for various minerals. For minerals of the same structure and different chemical composition (e.g., Fe content), the empirical relation $v_\phi \propto \bar{M}^{-1/2}$ is verified (after Chung 1972).

of the velocity with respect to the mean atomic mass is equal to $-\frac{1}{2}$, and hence equation (4.59) follows.

The relation (4.59) has been verified for various oxides (Chung 1972) (Fig. 4.6) and also for the geophysically interesting compounds with ilmenite (Liebermann 1982) and perovskite (Liebermann et al. 1977) structures (Fig. 4.7).

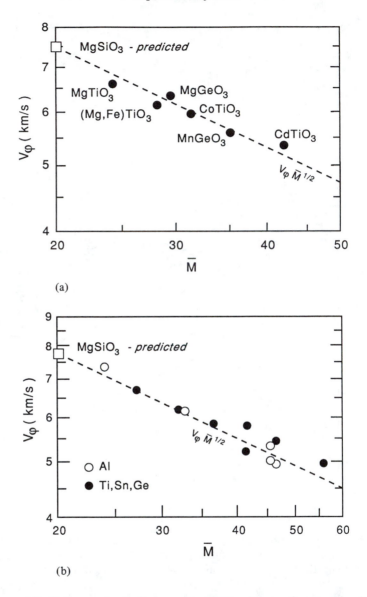

Figure 4.7. Bulk sound velocity versus mean atomic mass for crystals with ilmenite structure (a) and perovskite structure (b) (after Liebermann 1982).

Birch's postulate that, whatever the cause of the density change, the relation between v_Φ and ρ can be represented by a straight line of constant slope, must be somewhat qualified, although it remains valid as a general principle. It appears, for instance, that crystal structure has some effect on the coefficient b of the linear relation (4.55): The slope of the iso-$\bar{M}$ lines tend to be smaller ($b \leq 3$) for close-packed structures such as corundum, rocksalt, and perovskite than for open structures such as olivine, pyroxene, or quartz ($b \geq 4$) (Liebermann 1982). Furthermore, when the density change is caused by a phase transformation, b tends to be smaller when there is an increase in coordination than when the coordination remains unchanged (Liebermann and Ringwood 1973; Liebermann 1982).

A power-law form of the bulk-velocity–density relation can indeed be derived from equation (4.58), which we can write as

$$v_\Phi = a(\bar{M})\rho^{(\gamma - 1/3)} \tag{4.61}$$

D. L. Anderson (1967) proposed a power-law seismic equation of state, derived from the expression of pressure in terms of interatomic potentials (see Section 4.4). Following Gilvarry (1957a), Anderson writes equation (4.48) as:

$$P = (N-M)^{-1}K_0\left[\left(\frac{\rho}{\rho_0}\right)^N - \left(\frac{\rho}{\rho_0}\right)^M\right] \tag{4.62}$$

with $N = 1 + n/3$ and $M = 1 + m/3$. Taking the derivative of the pressure with respect to density and remembering that $\Phi = dP/d\rho$, we have

$$\Phi = \Phi_0(N-M)^{-1}\left[N\left(\frac{\rho}{\rho_0}\right)^{N-1} - M\left(\frac{\rho}{\rho_0}\right)^{M-1}\right] \tag{4.63}$$

and

$$\frac{d\ln\Phi}{d\ln\rho} = \frac{\rho}{\rho_0}\left[N(N-1)\left(\frac{\rho}{\rho_0}\right)^{N-2} - M(M-1)\left(\frac{\rho}{\rho_0}\right)^{M-2}\right]$$
$$\times\left[N\left(\frac{\rho}{\rho_0}\right)^{N-1} - M\left(\frac{\rho}{\rho_0}\right)^{M-1}\right]^{-1}$$

if $\rho/\rho_0 \approx 1$; that is, for small compressions, we have

$$\frac{d\ln\Phi}{d\ln\rho} \approx N + M - 1 = \frac{1}{3}(n+m+3) \tag{4.64}$$

or

$$\frac{\rho}{\rho_0} = \left(\frac{\Phi}{\Phi_0}\right)^{3/(n+m+3)} \tag{4.65}$$

Taking into account the expression of Slater's gamma (4.51), we can write

$$\frac{\rho}{\rho_0} = \left(\frac{\Phi}{\Phi_0}\right)^{1/[2(\gamma - 1/3)]} \tag{4.66}$$

which is identical to (4.61). Note that (4.66) can be written

$$\frac{d \ln \Phi}{d \ln \rho} = 2\left(\gamma - \frac{1}{3}\right) \tag{4.67}$$

For $N = 4 = K_0'$ and $M = 0$, equation (4.61) becomes Murnaghan's linear integrated EOS and (4.65) becomes

$$\frac{\rho}{\rho_0} = \left(\frac{\Phi}{\Phi_0}\right)^{1/3} \tag{4.68}$$

which, of course, can be directly obtained by taking the derivative of (4.6), with $K_0' = 4$,

$$\Phi = \frac{dP}{d\rho} = \Phi_0 \left(\frac{\rho}{\rho_0}\right)^{K_0' - 1} \tag{4.69}$$

For $N = 7/3$ and $M = 5/3$ ($n = 4$, $m = 2$), equation (4.62) becomes the Birch–Murnaghan EOS and we also have

$$\frac{\rho}{\rho_0} = \left(\frac{\Phi}{\Phi_0}\right)^{1/3} \tag{4.70}$$

which corresponds to a value of Slater's gamma of 1.83.

From an experimental correlation on 31 minerals and rocks, Anderson (1967) gives

$$\frac{\rho}{M} = 0.048 \Phi^{0.323} \tag{4.71}$$

in good agreement with the theoretical considerations (Fig. 4.8).

Equation (4.70) can be written

$$\ln \rho = \tfrac{1}{3} \ln \Phi + \text{const.} = \tfrac{1}{3}(\ln K - \ln \rho) + \text{const.}$$

or

$$\ln K = 4 \ln \rho + \text{const.} \tag{4.72}$$

Relation (4.72) can be directly derived by setting $dK/dP = 4$ in

$$\Phi = \frac{d \ln K}{d \ln \rho} = \frac{-d \ln K}{d \ln V} = \frac{dK}{dP}$$

Anderson and Nafe (1965) verified that in the case of oxides, the slope of the $\Phi - V$ logarithmic plot takes values between -4 and -3, whether the change in volume is due to compression or compositional variation. However, alkali halides, fluorides, sulfides, and covalent compounds behave differently and obey a relation

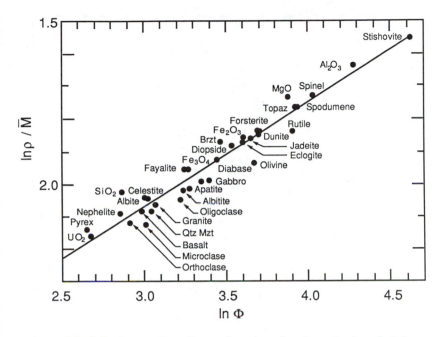

Figure 4.8. Seismic equation of state for selected rocks and minerals (after D. L. Anderson 1967).

$$K_0 V_0 = \text{const.} \qquad (4.73)$$

where the bulk modulus and the specific molar volume are taken at ambient pressure.

Anderson and Anderson (1970) and Chung (1972) showed that the relation (4.73) is indeed applicable to oxides, provided they have the same structure (Fig. 4.9). This results from the near constancy of the quantity $\alpha q^2 (n-1)/R_0$ in (4.60), which is equal to $K_0 V_0$ (see eq. (3.20)).

4.6 Thermal equations of state

So far, we have dealt only with isothermal equations of state, whose material parameters are experimentally determined at room temperature. We will now consider equations of state valid at high temperatures (above the Debye temperature) or with explicit temperature dependence.

i. At temperatures above the Debye temperature, we already know, of course, one appropriate EOS: the Mie–Grüneisen equation of state (3.53). However, it has the important drawback of containing a thermal

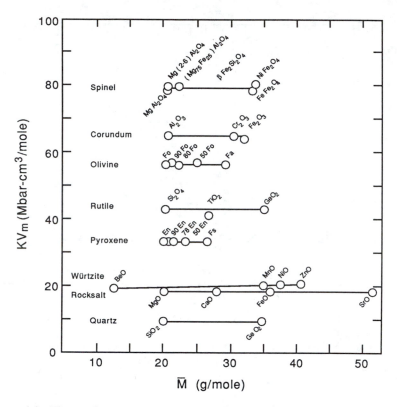

Figure 4.9. The product KV_m of the bulk modulus by the molar volume is almost constant for materials of the same crystal structure and independent of the mean atomic mass (after Chung 1972).

pressure term $P_{th} = \gamma E_{th}/V$, which is not directly accessible to experiment. O. L. Anderson (1984) circumvented the difficulty by proposing for P_{th} an empirical expression, independent of volume and linear in temperature: $P_{th} = a + bT$. The coefficients a and b are given a physical meaning by writing the thermal pressure along an isochore:

$$P_{th} = \int_0^T \left(\frac{\partial P}{\partial T}\right)_V dT = \int_0^T \alpha K_T\, dT$$

Assuming that αK_T is a constant independent of temperature for $T > \Theta_D$, an assumption valid in most cases (Birch 1968; Brennan and Stacey 1979; O. L. Anderson 1980), Anderson finds that

$$P_{th} = -\int_0^{\Theta_D} \alpha K_T\, dT + \alpha K_T (T - \Theta_D) \tag{4.74}$$

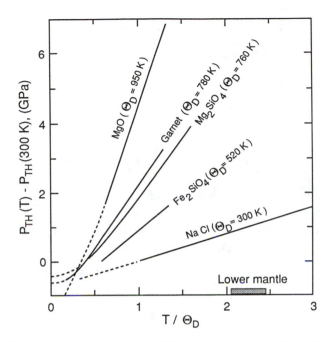

Figure 4.10. Experimentally determined thermal pressure (relative to the thermal pressure at room temperature) versus T/Θ_D ($\Theta_D =$ Debye temperature). In most cases the thermal pressure is proportional to temperature.

The Mie–Grüneisen EOS, thus modified, is claimed to be universal, since the linear correlation (4.74) holds for most materials (Fig. 4.10). The bulk modulus can be measured at high temperatures by resonance methods (Sumino, Anderson, and Suzuki 1983) and the thermal expansion coefficient α can be measured by dilatometry.

ii. Brennan and Stacey (1979) derived a high-temperature, thermodynamically based EOS by equating the expression of the thermodynamic gamma (3.55) with that of the Vashchenko–Zubarev gamma (3.81), both valid in the classical high-temperature range ($T > \Theta_D$). If $C_V =$ const. and $(dK_T/dT)_V = 0$, αK_T is independent of T and P (or ρ) and we can write

$$\gamma = \gamma_0 x^{-1}$$

with $x = \rho/\rho_0$. Denoting the derivatives with respect to x by primes, we have

$$K = xP' \tag{4.75}$$

$$\frac{dK}{dP} = 1 + \frac{xP''}{P'} \tag{4.76}$$

Inserting (4.75) and (4.76) into the expression of the Vashchenko–Zuba-rev gamma (3.81), Brennan and Stacey obtain a differential equation,

$$9x^3 P'' - (6x^2 + 18\gamma_0 x)P' + (4x + 24\gamma_0)P = 0 \qquad (4.77)$$

whose solution is the EOS:

$$P = \frac{K_0}{2\gamma_0}\left(\frac{\rho}{\rho_0}\right)^{4/3}\left\{\exp\left[2\gamma_0\left(1 - \frac{\rho}{\rho_0}\right)\right] - 1\right\} \qquad (4.78)$$

$$K = \frac{K_0}{3\gamma_0}\left(\frac{\rho}{\rho_0}\right)^{1/3}\left\{\left[2\frac{\rho}{\rho_0} + 3\gamma_0\right]\exp\left[2\gamma_0\left(1 - \frac{\rho}{\rho_0}\right)\right]^{-2}\frac{\rho}{\rho_0}\right\} \qquad (4.79)$$

For $P = 0$, $K_0' = 2\gamma_0 + \frac{5}{3}$, and for infinite pressure, K' asymptotically tends to $\frac{4}{3}$.

iii. Gilvarry (1957a) started from the isothermal general form of equations of state (4.47), which he wrote as

$$P = (N - M)^{-1}K_0\left[\left(\frac{V_0}{V}\right)^N - \left(\frac{V_0}{V}\right)^M\right] \qquad (4.80)$$

$$K = (N - M)^{-1}K_0\left[N\left(\frac{V_0}{V}\right)^N - M\left(\frac{V_0}{V}\right)^M\right] \qquad (4.81)$$

and he generalized it to arbitrary temperatures by replacing V_0 and K_0 by quantities $\mathcal{V}$ and $\mathcal{K}$, which are functions only of temperature. The temperature dependence of $\mathcal{V}$ and $\mathcal{K}$ is determined by considerations of thermodynamic consistency, demanding that the expressions of $(\partial P/\partial T)_V$ and $(\partial K/\partial T)_V$ obtained by differentiating (4.80) and (4.81) are equal to

$$\left(\frac{\partial P}{\partial T}\right)_V = \alpha K_T$$

and

$$\left(\frac{\partial K}{\partial T}\right)_V = -\alpha V\left(\frac{\partial K}{\partial V}\right)_T + K^2\left(\frac{\partial \alpha}{\partial P}\right)_T$$

After various reasonable approximations, Gilvarry finds:

$$\mathcal{V}(T) = V_0\exp\left[\int_0^T \alpha_0\, dT\right] \qquad (4.82a)$$

$$\mathcal{K}(T) = K_0\exp\left[\int_0^T \eta_0\alpha_0\, dT\right] \qquad (4.82b)$$

with $\eta = -K\alpha^{-1}(\partial\alpha/\partial P)_T$.

$\mathcal{V}(T)$ and $\mathcal{K}(T)$ are thus the volume of the solid at temperature T and zero pressure and the bulk modulus at temperature T and zero pressure, respectively. The equation of state obtained by replacing V_0 and K_0 by $\mathcal{V}(T)$ and $\mathcal{K}(T)$ in (4.80) can be approximated at high pressure by retaining only the first terms in the expansion of the exponentials:

$$P = P(T_0) + (N - M)^{-1} K_0 \left[N \left(\frac{V_0}{V} \right)^N - M \left(\frac{V_0}{V} \right)^M \right] \int_{T_0}^{T} \alpha_0 \, dT \qquad (4.83)$$

Gilvarry found a very good agreement between the pressure as a function of compression measured for potassium at 77 K and the prediction of his EOS based on experimental results at 4.2 K.

4.7 Shock-wave equations of state

4.7.1 Generalities

Although pressures up to above 5 Mbar seem to have been obtained in diamond–anvil cells (Xu, Mao, and Bell 1986), in practice, equations of state of minerals obtained by static compression using the diamond–anvil cell cover a pressure range seldom extending above 1 Mbar. For the measurement of the density as a function of pressure, from 1 bar to 3.7 Mbar (pressure at the center of the Earth), the shock-wave technique still has no competitors. Shock waves are generated in target samples by a variety of methods, essentially using explosives or high-velocity projectiles launched by a light-gas (hydrogen) gun. The quantities measured are the shock-wave velocity and the "particle velocity" (i.e., the velocity imparted to the particles of the sample by the shock wave) as well as the temperature. Details on the experimental apparatus and methods of measurement are given in recent review articles (Ahrens 1980, 1987).

The results of the shock-wave experiments are embodied in *Hugoniot curves* or "Hugoniots," loci of the peak shock states achieved from the initial state by experiments with different impact velocities. The Hugoniots are usually given as pressure versus density curves or sometimes as pressure versus volume curves. Typically, they exhibit several stages (Fig. 4.11).

i. Up to the pressure of the *Hugoniot elastic limit* (HEL), which can vary from about 2 kbar (for halite) to about 200 kbar (for diamond), the regime is one of finite elastic strain, corresponding to the propagation of the longitudinal shock wave.

ii. Above the Hugoniot elastic limit, the material plastically yields at the microscopic level and loses most or all of its shear strength, behaving in effect hydrostatically like a fluid. The Hugoniot then differs little, if at all, from the theoretical hydrostatic Hugoniot or "hydrostat."

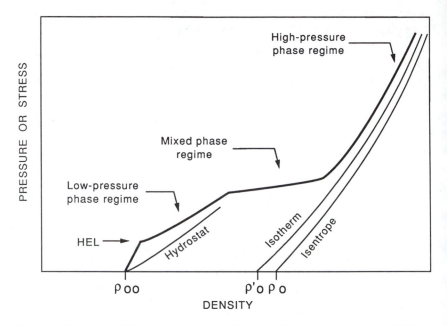

Figure 4.11. Typical Hugoniot curve showing the Hugoniot elastic limit (HEL). The hydrostatic compression curve (hydrostat) of the low-pressure phase is given for comparison (after Ahrens 1980).

iii. Most earth materials at high pressure exhibit one or several phase transitions toward higher-density phases, giving rise to new stages on the Hugoniot corresponding to the behavior of the high-pressure phases. Between the low-pressure phase and the high-pressure phase regimes (and between successive high-pressure phase regimes), a transitional "mixed-phase regime" occurs.

It is important to note that the Hugoniot is just the locus of the final shock state achieved, but that it is not the thermodynamic path followed by the material. In other words, "successive shock states along the Hugoniot cannot be achieved one from another by the shock process" (Ahrens 1987). The actual thermodynamic path followed is a straight line from the initial to the final state, called the *Rayleigh line*. Similarly, the pressure–density relation represented by the Hugoniot in the low- or high-pressure phase regimes does not correspond to an isothermal equation of state, any more than it is adiabatic. Isotherms or isentropes (adiabats) can, however, be derived from the Hugoniot (reduction of the Hugoniot to isothermal EOS) (see Sec. 4.7.3).

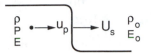

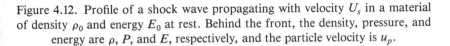

Figure 4.12. Profile of a shock wave propagating with velocity U_s in a material of density ρ_0 and energy E_0 at rest. Behind the front, the density, pressure, and energy are ρ, P, and E, respectively, and the particle velocity is u_p.

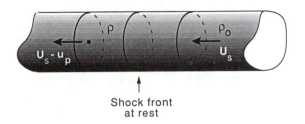

Shock front
at rest

Figure 4.13. Velocities of the particles in the frame of reference of the shock front.

We will now derive the Rankine–Hugoniot equations governing the thermodynamics of shock-wave experiments.

4.7.2 The Rankine–Hugoniot equations

Let us consider a steady planar shock front, with a rise time short compared to the characteristic decay time, propagating in a solid at rest (Fig. 4.12). The shock front propagates with a velocity U_s with respect to the frame of reference of the laboratory and imparts a velocity u_p to the particles of the solid. The specific mass and the internal energy per unit mass change from the values ρ_0 and E_0 at rest to ρ and E behind the shock front.

We will now perform the calculations in the frame of reference of the shock front, considered as at rest.

Let us assume that the sample is a cylinder of unit area cross section at rest in the frame of reference of the laboratory and that the shock wave goes through it from left to right with a velocity U_s (Fig. 4.13).

In the frame of reference of the shock front, the particles of matter arrive from the right with a velocity U_s (from right to left) and they acquire a velocity u_p (from left to right), the velocity with respect to the shock front therefore becomes $U_s - u_p > 0$ (the positive sense is taken from right to left).

Let us now write the conservation of mass, momentum, and energy in a slice containing the shock front.

i. Per unit time, the mass entering the slice is equal to the mass leaving it:

$$\rho_0 U_s = \rho(U_s - u_p) = m \tag{4.84}$$

ii. The sum of the forces exerted on the slice is equal to the rate of change of momentum. The forces on the unit-area-cross-section ends of the slice are equal to the normal stresses σ_0 and σ, that is, to the pressures P_0 and P in the hydrostatic case, above the HEL:

$$m[(U_s - u_p) - U_s] = \sigma_0 - \sigma$$

or

$$P - P_0 = m u_p \tag{4.85}$$

and with (4.84),

$$P - P_0 = \rho_0 U_s u_p \tag{4.86}$$

iii. The work done by the forces is equal to the sum of the increments of kinetic energy and internal energy per unit time

$$P_0 U_s - P(U_s - u_p) = m[E - E_0 + \tfrac{1}{2}(U_s - u_p)^2 - \tfrac{1}{2}U_s^2]$$

or, with (4.85),

$$-m u_p U_s + P_0 u_p + m u_p^2 = m[E - E_0 - u_p U_s + \tfrac{1}{2}u_p^2]$$

and, with (4.84),

$$\rho_0 U_s(E - E_0 - \tfrac{1}{2}u_p^2) = P_0 u_p \tag{4.87}$$

We can set $P_0 = 0$ in (4.84), (4.86), and (4.87) and obtain the usual form of the Rankine–Hugoniot equations

$$\rho = \frac{\rho_0 U_s}{U_s - u_p} \tag{4.88}$$

$$P = \rho_0 u_p U_s \tag{4.89}$$

$$\Delta E = E - E_0 = \tfrac{1}{2}u_p^2 \tag{4.90}$$

We see that the increase in internal energy is equal to the kinetic energy (per unit mass).

If, now, we retain P_0 for symmetry and eliminate u_p between (4.84) and (4.86), we obtain

$$U_s^2 = (P - P_0)\left[\frac{\rho}{\rho_0}(\rho - \rho_0)\right]$$

or, in terms of the specific volumes,

$$U_s = V_0 \left(\frac{P - P_0}{V_0 - V} \right)^{1/2} \tag{4.91}$$

Eliminating U_s between (4.86) and (4.91), we obtain

$$u_p = [(P - P_0)(V_0 - V)]^{1/2} \tag{4.92}$$

With (4.91) and (4.92), (4.87) gives

$$E = E_0 + \tfrac{1}{2}(P + P_0)(V_0 - V) \tag{4.93}$$

For a final shock state characterized by u_p and U_s, the energy is known, since it is a function of the shock-wave and particle velocities only, and the pressure and specific volume are, from (4.91) and (4.92),

$$P = P_0 + \frac{u_p U_s}{V_0} \tag{4.94}$$

$$V = V_0 - V_0 \frac{u_p}{U_s} \tag{4.95}$$

The initial specific volume is that of the sample, equal to that of the material for nonporous samples only. If an artificially porous sample is used (V_0 large), shock states of higher energy, at higher pressures, can be reached for the same shock-wave velocity. For a given specific volume (or mass), one can achieve different energy states (at different pressures) by driving shock waves into samples of different initial porosities. The Grüneisen parameter can then be obtained by carrying the finite differences ΔP and ΔE obtained from (4.94) and (4.93) into definition (2.60): $\gamma = V(\Delta P / \Delta E)_V$.

The Hugoniot $P(V)$ or $P(\rho)$ (Fig. 4.14) is, as seen above, the locus of the final shock states of given internal energy, which can be reached by following not the Hugoniot, but a straight line, the Rayleigh line.

For a constant wave velocity U_s and starting from $P_0 = 0$, we have, from (4.94),

$$P = \frac{u_p U_s}{V_0}$$

and, from (4.84),

$$u_p = \frac{U_s}{V_0}(V_0 - V)$$

Hence

$$P = \frac{U_s^2}{V_0^2}(V_0 - V) \tag{4.96}$$

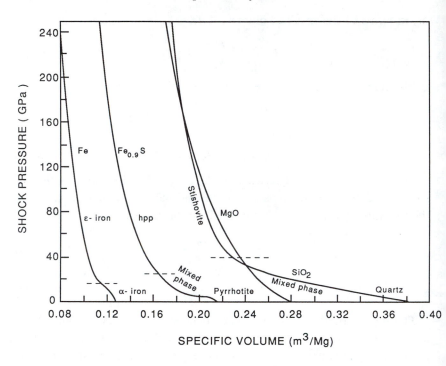

Figure 4.14. Pressure–volume Hugoniots for periclase, iron, pyrrhotite, and quartz. Except periclase, these minerals exhibit phase changes (after Ahrens 1987).

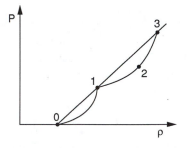

Figure 4.15. Multiple shock waves occur if the Rayleigh line from initial to final state intersects the Hugoniot. Point 1 corresponds to a phase transition or to the Hugoniot elastic limit. To obtain a final state between 1 and 3 (e.g., 2) two shock waves will form, one leading to the intermediate shock state 1, followed by a slower one leading from 1 to 2.

The slope of the Rayleigh line is equal to $-U_s^2/V_0^2$. It depends on the shock velocity and the initial specific volume only.

The Hugoniots generally exhibit a change in slope for the pressure corresponding to the Hugoniot elastic limit (Fig. 4.15). The states between the HEL and the intersection of the HEL Rayleigh line with the Hugoniot can be reached only by a bifurcation of the shock wave: Two successive shock waves form; the first one is an elastic wave that brings the material to the HEL state, the second one is slower and brings the material to the final state (Ahrens 1971, 1987).

After the passage of the shock wave, the shocked material is released from the high-pressure shock state to ambient pressure in a very short time (10^{-6} to 10^{-5} s), following a *release isentrope* (Ahrens 1987).

4.7.3 Reduction of the Hugoniot data to isothermal equation of state

For most materials, in a range of pressure where there is no phase transition, there exists an empirical linear relation between U_s and u_p, usually written (Takeuchi and Kanamori 1966)

$$U_s = C_0 + \lambda u_p \qquad (4.97)$$

Although there is no satisfactory theoretical explanation for this relation, Berger and Joigneau (1959), and later Ruoff (1965), have shown that, by combining the Rankine–Hugoniot equations, the Mie–Grüneisen equation of state, and an expansion of the pressure in a series of powers of the compression, it is possible to construct an expansion of U_s in a series of powers of u_p, whose first-order terms represent (4.97). The particle velocity tends toward zero with pressure and the wave velocity is then equal to the bulk sound velocity, hence

$$C_0 = \left(\frac{K}{\rho}\right)^{1/2} = \Phi^{1/2} \qquad (4.98)$$

The slope of the line is found to be

$$\lambda \approx \frac{(dK/dP)_0 + 1}{4} = \frac{\gamma_{DM} + 1}{2} \qquad (4.99)$$

With (4.97), (4.91), and (4.92), we obtain the Hugoniot centered at the zero-pressure state (metastable state, if relation (4.97) corresponds to a high-pressure phase) (McQueen et al. 1963)

$$P = P_0 + C_0^2 \frac{V - V_0}{[V_0 - \lambda(V_0 - V)]^2} \qquad (4.100)$$

where P is the pressure at high temperature under the shock conditions.

Raw Hugoniot data can also be fitted with a two-parameter Birch–Murnaghan-type equation constrained by choosing a zero-pressure density related to the slope of the $P(\rho)$ Hugoniot at zero pressure ($dP/d\rho = \Phi_0$) by the Anderson (1967) seismic equation of state (Anderson and Kanamori 1968; Ahrens, Anderson, and Ringwood 1969).

It is then necessary to remove the thermal pressure to obtain an isothermal equation of state at $T = 0$ K.

Takeuchi and Kanamori (1966) use the Mie–Grüneisen equation of state

$$P - P_K = \gamma \frac{E - E_K}{V} \qquad (4.101)$$

where P_K is the pressure needed to compress the material at 0 K to the same specific volume as that obtained under shock and E_K is the internal energy corresponding to the isothermal compression at 0 K. The isothermal equation of state relates P_K to V and V_{K_0} (specific volume at 0 K and zero pressure). It is found by integrating a system of differential equations essentially obtained by carrying (4.93) into the expression of γ from (4.10) and identifying it with Slater's or Dugdale–McDonald's expressions in terms of the derivatives of pressure and by using the definition $P_K = (\partial E_K / \partial V)_T$. The specific volume at 0 K is estimated using the definition of the thermodynamic gamma in terms of the coefficient of thermal expansion.

Shapiro and Knopoff (1969) proposed a method based on the same principles, but which they claimed to be mathematically simpler.

4.8 First principles equations of state

4.8.1 Thomas–Fermi equation of state

The Thomas–Fermi equation of state (TF EOS) is an EOS in which the atomic structure of the solid is not taken into consideration; the pressure is assumed to be entirely that of a degenerate electron gas and is calculated using the semi-classical Thomas–Fermi approximation. These assumptions are justified for elements at extremely high pressures, inside stars for instance. Even the pressure at the center of the Earth is much below the domain of validity of the TF EOS, which does not extend below

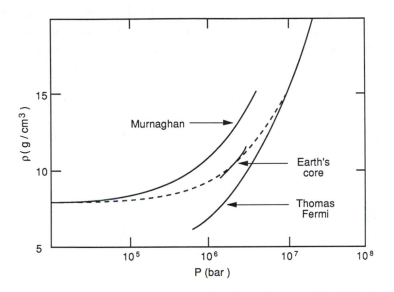

Figure 4.16. Estimated density of iron as a function of pressure. The dashed curve is interpolated to pass through the data for the core (after Birch 1952).

10 Mbar. This, apparently, should suffice to confine the interest in the TF EOS to astrophysics and exclude it from geophysics. It turns out however that considerable attention was given in the geophysical literature to the TF EOS (see, e.g., Boschi and Caputo 1969), extrapolating it below its validity range down to the range of shock-wave and static pressure experimental data. Although these attempts generally met with indifferent success, interpolation between the TF and experimental *PV* curves was possible (Elsasser 1951) and it provided a useful constraint on the composition of the core (Birch 1952) (Fig. 4.16). It is therefore justified to devote some space here to outlining the basic physics of the TF EOS.

i. The Thomas–Fermi approximation

The Thomas–Fermi approximation is a method for finding the density of electrons in an atom of atomic number Z. It essentially rests on the following assumptions (see Eliezer et al. 1986):

The system as a whole is in its lowest quantum state.

In a volume so small that within it the change in the potential energy is small compared with the mean total energy of an electron, the number of electrons is large.

It is therefore possible to consider that the free electrons form a gas that can be treated classically rather than by quantum mechanical methods. The only contribution of quantum mechanics is that the electrons are subject to the exclusion principle, and hence obey Fermi–Dirac statistics (Fermi–Dirac gas). The approximation works better for heavy metals.

Let the electrostatic potential due to the nucleus and electrons inside the atom be $V(\mathbf{r})$, $\mathbf{r}$ being the distance from the nucleus, and let the maximum energy of an electron be $E_{max} = -eV_0$ (e is the charge of the electron). At any point, we have

$$\frac{p^2}{2m} - eV \le E_{max}$$

where p is the momentum of an electron, whose maximum value is

$$p_{max} = [2me]^{1/2}[V(r) - V_0]^{1/2} \tag{4.102}$$

According to Fermi–Dirac statistics, each volume of phase space $d\mathbf{p}d\mathbf{r}$ contains $(2/h^3)d\mathbf{p}d\mathbf{r}$ electrons and the volume of the momentum space corresponding to points which are occupied is $(4\pi/3)p_{max}^3$, hence the number of electrons per unit volume in real space is

$$n(\mathbf{r}) = \frac{8\pi}{3h^3}(2me)^{3/2}[V(r) - V_0]^{3/2} \tag{4.103}$$

The electrostatic potential $V(r)$ and the density of charge $n(r)$ are related by Poisson's equation

$$\nabla^2 V = 4\pi ne \tag{4.104}$$

With spherical symmetry, the Laplacian is written:

$$\nabla^2 V = \frac{1}{r^2}\frac{d}{dr}\left(r^2\frac{dV}{dr}\right) \tag{4.105}$$

As $r \to 0$, the potential is due mostly to the nucleus and behaves as Ze/r. It is therefore convenient to introduce the dimensionless quantity

$$\Phi(r) = [V(r) - V_0]\left(\frac{Ze}{r}\right)^{-1} \tag{4.106}$$

with the boundary condition $\Phi(0) = 1$. We have therefore

$$\nabla^2 V = \frac{Ze}{r}\frac{d^2\Phi}{dr^2} \tag{4.107}$$

With (4.107) and (4.103), we can write (4.104) as

$$\frac{d^2\Phi}{dr^2} = \frac{32\pi^2}{3h^3}(2me^2)^{3/2}Z^{1/2}\Phi(r)^{3/2}r^{-1/2} \tag{4.108}$$

Introducing the dimensionless variable $x = r/\mu$, with $\mu \propto Z^{-1/3}$ (see Eliezer et al. 1986), we obtain the dimensionless Thomas–Fermi differential equation, valid for any value of Z,

$$\frac{d^2\Phi}{dx^2} = \Phi^{3/2}x^{-1/2} \tag{4.109}$$

The equation can be solved numerically (Feynman, Metropolis, and Teller 1949), with $\Phi(x)$ expanded in a series of powers of x:

$$\Phi = 1 + a_2 x + a_3 x^{3/2} + a_4 x^4 + \cdots \tag{4.110}$$

The coefficients are expressed as functions of a_2. The values of Φ are tabulated as a function of $x_0 = r_0/\mu(Z)$, where r_0 is the atomic radius of the element of atomic number Z.

ii. Pressure–volume relation at $T = 0$ K

From (4.103) and (4.106), the electronic density of an isolated atom is

$$n = \frac{8\pi}{3h^3}(2me)^{3/2}\left[\frac{Ze}{r}\Phi\right]^{3/2} \tag{4.111}$$

The solid is considered as a gas of atoms, each atom being surrounded by a sphere of atomic size; the pressure is due to the bombardment of the free electrons on the boundary of the atomic sphere and is given by the kinetic theory as

$$P = \frac{2}{3}\frac{E_0}{V} = \frac{2}{3}\frac{E_0 n}{N} \tag{4.112}$$

$n = N/V$ is the number of free electrons per unit volume and E_0/N is the internal energy (equal to the kinetic energy) per electron:

$$\frac{E_0}{N} = \frac{3h^2}{10m}\left(\frac{3}{8\pi}\right)^{2/3}n^{2/3} \tag{4.113}$$

Hence

$$P = \frac{h^2}{5m}\left(\frac{3}{8\pi}\right)^{2/3}n^{5/3} \tag{4.114}$$

With (4.114) and (4.111) and remembering that $r = \mu x$, we have

$$PV = \frac{2}{15}\frac{Z^2 e^2}{\mu}x_0^{1/2}[\Phi(x_0)]^{5/2} \tag{4.115}$$

where V is the atomic volume

$$V = \frac{4\pi}{3}(\mu x_0)^3 \tag{4.116}$$

The values of $\Phi(x_0)$ for a given atomic number are found in tables (Feynman et al. 1949).

At high enough pressures, for an element of atomic number Z, the pressure can be directly found from (4.114), by taking for n the number of electrons in the atomic volume, $n = Z/V$:

$$P = \frac{h^2}{5m} \left(\frac{3}{8\pi} \right)^{2/3} Z^{5/3} V^{-5/3} \tag{4.117}$$

or

$$PV = \frac{h^2}{5m} \left(\frac{3}{8\pi} \right)^{2/3} Z^{5/3} V^{-2/3} \tag{4.118}$$

Brillouin (1954, p. 245) estimated the electronic pressure for metals by taking one free electron per atom; he found values of the order of 100 kbar.

Scaled pressure–volume curves, valid for all values of Z, are often given as $PZ^{-10/3}$ versus ZV. This can be justified by writing (4.115) as

$$P \propto Z^2 \mu^{-4} \left[\frac{\Phi(x_0)}{x_0} \right]^{5/2} \tag{4.119}$$

Remembering that $\mu \propto Z^{-1/3}$, we see from (4.116) that x_0 is a function only of ZV and we can write

$$PZ^{-10/3} \propto \left[\frac{\Phi(x_0)}{x_0} \right]^{5/2} = f(ZV) \tag{4.120}$$

Dirac modified the Thomas–Fermi theory to include the quantum exchange effects (TFD model). The resulting equation cannot be put in a scaled form and has to be solved for every value of Z (see Eliezer et al. 1986; Boschi and Caputo 1969). The results for iron, compared with the isotherm derived from shock-wave experiments and the Birch–Murnaghan EOS (Takeuchi and Kanamori 1966), are given in Figure 4.17.

Feynman et al. (1949) treated the more complicated case of the TF EOS at high temperature.

4.8.2 Ab initio quantum mechanical equations of state

Construction of parameter-free, ab initio equations of state from quantum mechanics for specific minerals is a rapidly developing field and it is beyond the scope of this book to enter into the quantum mechanical details of the methods. We will only and very briefly give the physical basis of the more popular models and quote a few recent publications where the relevant references can be found.

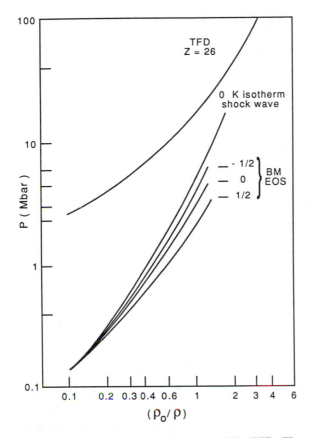

Figure 4.17. Pressure–density isotherms for iron at 0 K. TFD: Thomas–Fermi–Dirac approximation; BM: Birch–Murnaghan equation of state for $\xi = -\frac{1}{2}, 0, \frac{1}{2}$ (after Boschi and Caputo 1969).

Hemley, Jackson, and Gordon (1985) calculated the binding energy of the crystal $U(\rho)$, where ρ is the total charge density, as the sum of several contributions:

$$U(\rho) = E_{LRE}(\rho) + E_{SRE}(\rho) + E_{KE}(\rho) + E_{EX}(\rho) + E_{COR}(\rho) \qquad (4.121)$$

$E_{LRE}(\rho)$ is the long-range electrostatic energy calculated by the Ewald summation method, $E_{SRE}(\rho)$ is the short-range electrostatic energy, $E_{KE}(\rho)$ is the kinetic energy, $E_{EX}(\rho)$ is the exchange energy, and $E_{COR}(\rho)$ is the correlation energy; the last four terms are calculated using the density functionals of the Modified Electron Gas (MEG) theory, in which the ions are assumed to behave rigidly. The pressure as a function of volume and temperature is

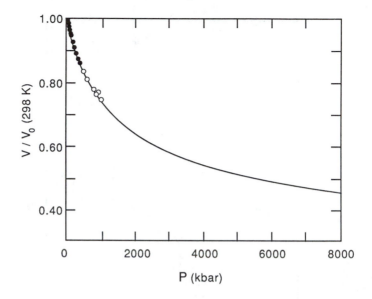

Figure 4.18. Calculated isothermal compression curve for MgO at 298 K. The circles represent experimental results (after Hemley et al. 1985).

$$P = P_E + P_{ZP} + P_{TH} \qquad (4.122)$$

where P is the applied pressure, $P_E = -\partial U(\rho)/\partial V$, the electronic pressure calculated from (4.121), P_{ZP} is the zero-point pressure, and P_{TH} is the thermal pressure. The last two terms are calculated from lattice dynamics in the quasi-harmonic approximation. The resulting EOS, calculated for MgO, with no parameter adjusted to fit experimental data, is in good agreement with the experimental EOS (Mao and Bell 1979) (Fig. 4.18).

Hemley, Bell, and Mao (1987) and Wolf and Bukowinski (1985, 1987) used the same method to calculate the temperature-dependent EOS of $MgSiO_3$ and $CaSiO_3$ perovskites; the structures were determined by energy minimization.

Cohen (1987) also calculated the EOS of $MgSiO_3$ perovskite and MgO, using the potential induced breathing (PIB) model, which allows the ions to behave in a nonrigid fashion. The resulting EOS is in excellent agreement with the experimental results of Yagi, Mao, and Bell (1982) and Knittle and Jeanloz (1987) (Fig. 4.19). The parameters K_0 and K_0' of the equation of state at room temperature of $MgSiO_3$ perovskite, found by analogue prediction, experiments, and ab initio calculations are gathered in Table 4.2.

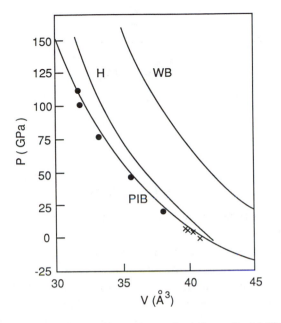

Figure 4.19. Calculated and experimental equation of state for MgSiO₃ perovskite at room temperature. WB: Wolf and Bukowinski 1987; H: Hemley, Jackson, and Gordon 1987; PIB: potential induced breathing (Cohen 1987). The experimental results of Yagi et al. (1982) and Knittle and Jeanloz (1987) are indicated by crosses and circles, respectively (after Cohen 1987).

Table 4.2. *Parameters of the equation of state at room temperature for MgSiO₃ perovskite*

Reference	K_0 (kbar)	K_0'	Method
Liebermann et al. (1977)	250	—	Analogue
Yagi et al. (1982)	260	3–5	X rays
Knittle and Jeanloz (1987)	266	3.9	X rays
Kudoh et al. (1987)	247	4.0	X rays
Yeganeh-Haeri, Weidner, and Ito (1989)	246	—	Brillouin
Hemley et al. (1987)	335	0.2	MEG
Wolf and Bukowinski (1985)	260	4.0	MEG
Cohen (1987)	247	4.1	PIB

5

Melting

5.1 Generalities

Melting is an extremely important phenomenon for solids since it causes them to cease being solids and to transform to the liquid state of matter, thereby losing crystalline long-range order and resistance to shear. Melting is a first-order phase transition, that is, it exhibits discontinuities in the first derivatives of the free energy: volume and entropy.

Pure metals (and more generally elements) have a fixed melting point T_m at a given pressure. The equilibrium between solid and liquid is univariant and the variation of the melting point with pressure is represented by the *melting curve* $T_m(P)$.

The situation is more complicated for multicomponent systems such as alloys and mineral crystals. Several cases may arise and are best understood by referring to phase diagrams (see also Sec. 7.4). A *phase diagram* at constant pressure is a map of the stability domains of the various possible phases in the composition–temperature space. For the sake of simplicity, let us consider only binary systems, whose composition can be defined by the proportions (in atomic or mole percents) of two simple end-members. The end-members that will concern us here can be elements (e.g., Fe, S) or simple binary oxides (e.g., MgO, SiO_2) that exhibit *congruent melting* like elements (i.e., the solid melts at fixed temperature, giving a liquid with the same chemical composition). For various definite proportions of the end-members, compounds may exist, for example, FeS, FeS_2, or MgO, SiO_2 ($MgSiO_3$) or $2MgO$, SiO_2 (Mg_2SiO_4). The compounds may sometimes in turn be used as end-members in other, more restricted phase diagrams, for example, $MgSiO_3$–Mg_2SiO_4.

The most typical melting situations are schematically represented in Figure 5.1:

i. Close to one end-member (e.g., α), there usually is a region where the stable phase is a solid solution of β in α. For a composition x_1 (Fig.

96

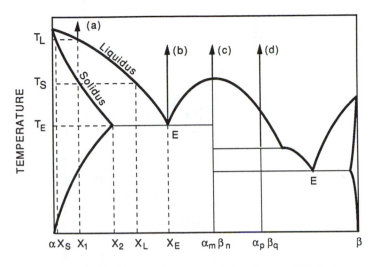

Figure 5.1. Typical phase diagram for compositions intermediate between α and β. (α, β in the cases that interest us are often simple oxides.) (a) For compositions such as x_1, when temperature T_s is reached during heating (cooling) the first (last) drop of liquid of composition x_1 appears (disappears); when temperature T_1 is reached during heating (cooling) the last (first) crystal of composition x_s disappears (appears). (b) For the eutectic composition x_E, the solid (liquid) melts (freezes) at the fixed temperature T_E of the eutectic point E. (c) For the definite composition $\alpha_m \beta_n$, the solid melts congruently, yielding a liquid of the same composition. (d) For the definite composition $\alpha_p \beta_q$, the solid melts incongruently, decomposing into a liquid in equilibrium with crystals of composition $\alpha_m \beta_n$.

5.1a) melting occurs over a temperature interval, with a variable composition of the liquid: When the temperature reaches the value T_S, the first drop of liquid of composition x_L appears, in equilibrium with a solid solution of composition x_1; if the temperature is increased, the proportion of liquid increases and its composition varies until the temperature T_L is reached, at which the last crystal of composition x_S disappears.

The locus of the points in the diagram corresponding to the appearance of the first drop of liquid is called *liquidus* and the locus of points corresponding to the disappearance of the last crystals (or appearance of first crystals during cooling) is called *solidus*.

The situation in Figure 5.1a corresponds to a rather usual case, where the melting point of a crystal is lowered by impurity elements in solution. A geophysically important example is that of the lowering of the melting point of iron by sulfur (Fig. 5.2). When end-members differ only by the chemical nature of atoms of comparable size on the same crystalline sites

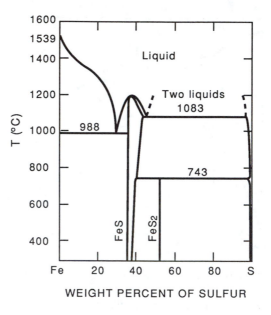

Figure 5.2. Phase diagram for the iron–sulfur system at ambient pressure (after
Verhoogen 1980).

(e.g., forsterite Mg_2SiO_4 and fayalite Fe_2SiO_4), a continuous solid solution can exist and the spindle formed by the solidus and liquidus extends
over the whole composition range (Fig. 5.3).

ii. A common feature of phase diagrams is the existence of one or several compositions for which melting occurs at a fixed temperature. For
these compositions, the liquidus exhibits an angular point corresponding
to a deep minimum of the melting temperature (*eutectic point*), as shown
in Figure 5.1b.

iii. Intermediate compounds may melt congruently (Fig. 5.1c) and can
be considered as end-members, thus dividing the phase diagram into independent binary diagrams; such is, for instance, the case of forsterite
Mg_2SiO_4 in the MgO–SiO_2 diagram (Fig. 5.4). Others melt, also at a fixed
temperature, but incongruently (Fig. 5.1d), decomposing into a liquid
and another compound; such is the case of enstatite $MgSiO_3$ at atmospheric pressure, which gives forsterite and a silica-rich liquid (Fig. 5.4).
At pressures higher than 5 kbar, melting becomes congruent and a eutectic
appears between forsterite and enstatite. The temperature of the eutectic point increases with pressure, in parallel with the melting point of

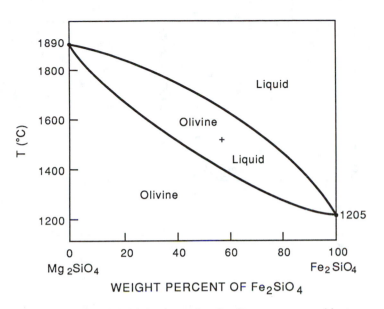

Figure 5.3. Phase diagram of the forsterite–fayalite system at ambient pressure.

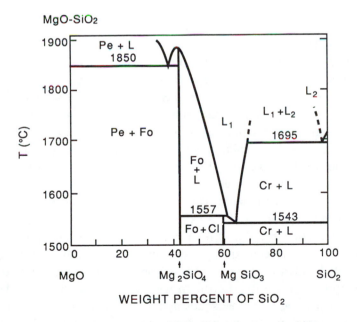

Figure 5.4. Phase diagram of the MgO–SiO$_2$ system at ambient pressure (after Levin et al. 1964). Fo: Forsterite, Cl: Clinopyroxene, Cr: Cristobalite, Pe: Periclase.

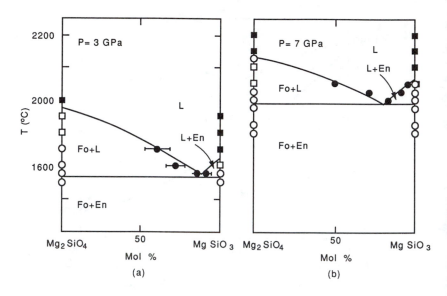

Figure 5.5. Phase diagram of the system forsterite–enstatite at pressures of 3 GPa (a) and 7 GPa (b) (after Kato and Kumazawa 1985b).

enstatite, and its composition shifts toward the Mg-rich side (Kato and Kumazawa (1985b) (Fig. 5.5).

Melting of rocks is even more complicated: Minerals with the lowest melting point melt first when temperature increases, and the liquid reacts with other minerals. To a given temperature above the melting point of the most fusible mineral corresponds a given degree of *partial melting*. One still can define a liquidus, on which the first fractions of melt are in equilibrium with the "liquidus phases"; see, for example, Takahashi (1986) and Ito and Takahashi (1987) on the melting of peridotites at pressures of the upper and lower mantle.

Melting in the Earth obviously concerns rocks, and melting of rocks is an important research topic in experimental petrology; it is however a complicated phenomenon, little amenable to physical extrapolations to high pressures. As a consequence, geophysicists generally use the melting curves of the mantle minerals and of iron to obtain constraints on the temperature profile in the Earth. In what follows, we will consider the physics of melting of simple solids as a basis for studying the effect of pressure on melting, and we will review the various melting laws giving T_m as a function of pressure.

5.2 Thermodynamics of melting

5.2.1 Clausius–Clapeyron relation

Let us express the equilibrium between solid and liquid at the melting point by writing that the total differential change in Gibbs free energy is equal to zero when a small volume element of the solid changes to liquid:

$$dG = (V_L - V_S)dP - (S_L - S_S)dT = 0 \qquad (5.1)$$

V_L, S_L and V_S, S_S are the specific volume and entropy per unit mass of the liquid and solid, respectively. The slope of the melting curve is given by the Clausius–Clapeyron equation:

▶
$$\frac{dT_m}{dP} = \frac{\Delta V_m}{\Delta S_m} \qquad (5.2)$$

where $\Delta V_m = V_L - V_S$ is the *melting volume* at pressure P, and $\Delta S_m = S_L - S_S$ is the *melting entropy* at pressure P.

The latent heat of melting is:

$$L = T_m \Delta S_m \qquad (5.3)$$

5.2.2 Volume and entropy of melting

i. The melting volume is usually positive, since in most cases the specific volume of the liquid is greater than that of the solid. For metals, the ratio of the melting volume to the specific volume of the solid is of the order of a few percent, whereas it is an order of magnitude greater for alkali halides; the silicates have intermediate values (see Table 5.1). When the short-range structure of the liquid is very different from that of the solid, due to a change in the type of bonding on melting, the liquid can be denser than the solid ($\Delta V_m < 0$). This is, for instance, the case for water, which is denser than ice I_h, and silicon, which becomes metallic on melting. Since the melting entropy is always positive (the liquid is always more disordered than the solid), the melting curve of these substances has therefor a negative slope: The melting point decreases as pressure increases. In some cases (e.g., Rb, Ba, Eu), the melting curve may exhibit a maximum at high pressures (Kawai and Inokuti 1968).

ii. The entropy of melting is generally the sum of two terms:

$$\Delta S_m = \left(\frac{\partial S}{\partial V}\right)_T \Delta V_m + \Delta S_d \qquad (5.4)$$

Table 5.1. *Melting volume ΔV_m in cm^3/mol,*
melting entropy ΔS_m in J/mol K, and slope
of the melting curve at 1 bar for various crystals

Crystal	ΔV_m	$\dfrac{\Delta V_m}{V}$	ΔS_m	$\left(\dfrac{dT_m}{dP}\right)_0$
Na	0.60	0.025	7.10	8.5
Mg	0.46	0.041	9.43	7.2
Al	0.64	0.060	11.3	5.9
K	1.20	0.026	7.10	16.9
Fe	0.28	0.036	8.32	3.5
Cu	0.32	0.042	9.58	3.64
Ag	0.42	0.038	9.27	6.04
Pb	0.46	0.041	8.28	7.23
LiF	3.47	0.327	24.12	11.2
NaCl	7.01	0.238	26.04	23.8
KCl	7.02	0.173	25.4	26.7
KBr	7.72	0.165	25.3	38.0
RbI	8.05	0.126	24.0	15.0
Quartz	2.5	0.105	5.53	35.5
Forsterite	3.8	0.081	70	4.8
Fayalite	3.7	0.076	60.9	7.5
Pyrope	18.3	0.158	162	5.5
Enstatite	5.4	0.160	41.1	12.8
Diopside			82.7	13.2

Sources: The values of ΔV_m and ΔS_m are from Ubbelohde
(1978) for metals and from Jackson (1977) for alkali halides.
The values of ΔV_m for silicates and of ΔS_m for forsterite
are from Bottinga (1985). The values (averaged) of ΔS_m for
the other silicates are from Richet and Bottinga (1986). The
values of the slopes of the melting curves are from Jackson
(1977).

The first term corresponds to the entropy change due to the modification of the vibrational frequencies, following the isothermal expansion
on melting, and the second term is the entropy of disorder due to the
structural differences between the solid and the less well-ordered liquid.
The disorder is generally positional and can also be rotational if the structural units of the liquid have no spherical symmetry.

Oriani (1951) calculated the expansion contribution to the entropy for
metals, starting from Maxwell's relations

$$\left(\frac{\partial S}{\partial V}\right)_T = \left(\frac{\partial P}{\partial T}\right)_V = -\left(\frac{\partial P}{\partial V}\right)_T \left(\frac{\partial V}{\partial T}\right)_P = \alpha K_T \qquad (5.5)$$

The first term in (5.4) can therefore be written

$$\Delta S_V = \alpha K_T \Delta V_m \qquad (5.6)$$

or

$$\Delta S_V = \gamma C_V \frac{\Delta V_m}{V} \qquad (5.7)$$

where γ is the thermodynamic gamma and V is the specific volume of the solid.

The existence of a linear relation between the entropy and the volume of melting for various metals is implicit in Oriani's analysis and directly obtains if the entropy of disorder is independent of the metal and if it can be assumed that αK_T is a constant for a given solid (see Sec. 3.5). Indeed, Stishov et al. (1973) noticed that the function $\Delta S_m/(\Delta V_m/V)$ is the same for argon and for sodium and tends to $R \ln 2$ when $\Delta V_m/V$ vanishes. Lasocka (1975) further reported that, for a number of metals, the representative points on a diagram ΔS_m versus $\Delta V_m/V$ were scattered about a straight line passing through the point with ordinate $R \ln 2$ for $\Delta V_m = 0$. Tallon (1980), probably unaware of Oriani's work, again justified the relationship

$$\Delta S_m = R \ln 2 + \alpha K_T \Delta V_m \qquad (5.8)$$

He also showed that the relationship is verified for sodium along its melting curve, but he did not provide an explanation for the fact that the entropy of disorder is close to $R \ln 2$.

Rivier and Duffy (1982) showed that, for dense atomic liquids, the configurational entropy of disorder is, indeed, equal to $R \ln 2$, by identifying the configurations of the liquid with its topological degrees of freedom, linked to the existence of line defects in the liquid that are absent in the solid.

The linear relationship between entropy and volume of melting is apparently valid only for dense atomic liquids (metals, rare gases, etc.); it is not verified for alkali halides and silicates, probably due to unaccounted for contributions. In the case of silicates, for instance, the entropy of melting must comprise a compositional (mixing) term, which is larger for minerals with isolated SiO_4 tetrahedra such as olivine and garnet than for chain silicates such as enstatite and is, of course, zero for quartz (Stebbins, Carmichael, and Moret 1984).

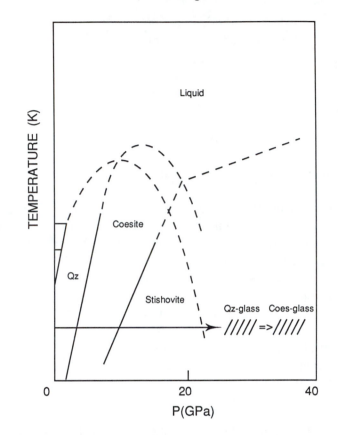

Figure 5.6. Phase diagram of SiO₂. The horizontal arrow indicates the path of pressure-induced amorphization (after Hemley et al. 1988).

Jackson (1977), searching for melting systematics, compared the melting curves of fluoride and fluoberyllate analogues of silicates (e.g., BeF_2, Li_2BeF_4). He found that the entropy of melting is primarily a function of the crystal structure, whereas the volume of melting is controlled by the molar volume of the crystal within each isostructural series. The magnitude of dT_m/dP is a function of the way in which the SiO_4 and BeF_4 tetrahedra are linked. It is small for structures with isolated tetrahedra and large for structures where the tetrahedra form a 3-D framework, the melting entropy being low in this case, due to the similarity between the solid and liquid phases.

Experimental values of ΔS_m, ΔV_m, and dT_m/dP are given in Table 5.1.

5.2.3 Metastable melting

For solids, such as ice, which decrease in volume upon melting ($\Delta V_m < 0$), the melting curve has a negative slope and usually ends at a triple point at high pressure, where another phase becomes stable. However, the melting curve of the low pressure phase can be extrapolated in the stability domain of the high-pressure phase. From this observation, Mishima, Calvert, and Whalley (1984) deduced that, if a low-pressure phase were pressurized at temperatures low enough to remain in a metastable state, it would melt when the trajectory of the representative point in the (P, T) plane crosses the extrapolated melting curve. Of course, for temperatures lower than the glass transition temperature, "melting" would produce not a liquid, but an amorphous glass. Mishima et al. (1984) performed the experiment on ice I at 77 K and succeeded in "melting" it at 10 kbar, producing a new high-density (1.31) amorphous phase of ice.

Hemley et al. (1988), compressing SiO_2 quartz and coesite at 300 K, transformed them into glass at 25–35 GPa, bypassing the stable higher-pressure phases, thus confirming that metastable melting can also occur in minerals (Fig. 5.6).

Richet (1988) observed that there is no reason why such "melting" (or rather vitrification) should be restricted to solids with $\Delta V_m < 0$ only; high-pressure phases with $\Delta V_m > 0$ should be decompressed rather than compressed to achieve vitrification. He gave a thermodynamic analysis of direct vitrification from the solid state and argued that the fact that high-pressure $CaSiO_3$ perovskite cannot be quenched and immediately vitrifies on release of pressure represents metastable melting.

5.3 Semi-empirical melting laws

5.3.1 Simon equation

The equation proposed by Simon and Glatzel (1929) satisfactorily describes the melting curve of many substances: solidified gases, metals, organic crystals, silicates. It was originally proposed in the form

$$\log(a + P) = c \log T + b$$

and tested for alkali metals and organic molecular compounds. Nowadays, it is usually written

$$\frac{P - P_0}{a} = \left(\frac{T_m}{T_0}\right)^c - 1 \tag{5.9}$$

Table 5.2. Parameters of Simon equation for a few substances

Substance	T_0 (K)	a (kbar)	c
Fe	1805	1070	1.76
Hg	234.3	382.15	1.177
Forsterite	2163	108.33	3.7
Fayalite	1490	157.80	1.59
Pyrope	2073	19.79	9.25
Enstatite	1830	28.65	5.01
Quartz	2003	15.99	3.34
Albite	1373	60.95	2.38

Note: $P_0 = 0$ for all substances, except for pyrope ($P_0 = 40$ kbar).
Sources: The parameters for Fe and Hg are from Babb (1963a); those for silicates are from Bottinga (1985).

where P_0 and T_0 are the pressure and temperature of the triple point and a and c are parameters that depend on the substance (they do not have the same meaning as in the original equation). For most substances, P_0 is close to zero and can be neglected before P. The slope at the triple point, close to the melting point at ambient pressure, is $dT_m/dP = T_0/ac$.

Several authors have shown that a melting equation of the same form as the Simon equation could be derived from physically based melting models: Domb (1951) started from the Lennard-Jones and Devonshire model and found that it led to a formula of the right type, at high enough pressure and for solids with central forces, but that the melting temperatures predicted were too high. Salter (1954) derived Simon's equation by eliminating the atomic volume between the Mie–Grüneisen equation of state and the Lindemann equation of melting. He found a relation between the constant c and the Grüneisen parameter, $c = (6\gamma + 1)/(6\gamma - 2)$. Gilvarry (1956c) (see also Babb 1963b) used the Lindemann law and the Murnaghan integrated linear equation of state to obtain an equation of the fusion curve similar to Simon's equation.

The Simon equation has been fitted to the melting curve of many substances (mostly organic crystals) by Babb (1963a) and for silicates by Bottinga (1985) (Table 5.2). It is convenient to analytically fit melting curves but, even though it has been given some theoretical justification, it cannot

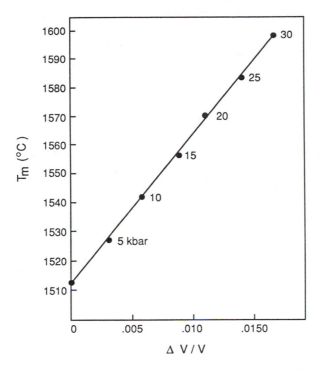

Figure 5.7. Melting temperature versus isothermal compression of iron at room
temperature (after Kraut and Kennedy 1966a, b).

be used to extrapolate the melting curve outside of the interval in which it
has been experimentally determined.

5.3.2 Kraut–Kennedy equation

Kraut and Kennedy (1966a, b) found that, for many substances and in a
wide pressure range, the melting temperature varies linearly as a function
of the compression of the solid $\Delta V/V_0$ (>0) or, in other words, the melting
curve is a straight line, if plotted against compression instead of pressure:

$$T_m = T_m^0 \left(1 + C\frac{\Delta V}{V_0}\right) \tag{5.10}$$

They claimed that the linear relation can be safely extrapolated up to a
maximum compression of 0.5. For iron (Fig. 5.7) Kraut and Kennedy find:

$$T_m(°C) = 1513\left(1 + 3.3209\frac{\Delta V}{V_0}\right) \tag{5.11}$$

The relation also holds for solids whose melting curve has a negative slope (e.g., germanium).

Gilvarry (1966) and Vaidya and Gopal (1966) independently showed that the Kraut–Kennedy relation can be derived from the Lindemann law (see Sec. 5.4.2) and that the constant C can be expressed as a function of the Grüneisen parameter at P_0, T_0: $C = 2(\gamma_0 - \frac{1}{3})$. The Kraut–Kennedy law can also be derived from the Clausius–Clapeyron relation, as shown by Libby (1966) and Mukherjee (1966). If the pressure dependence of the bulk modulus is neglected and if $\Delta V_m / \Delta S_m$ is assumed to remain constant along the melting curve, integration of the Clausius–Clapeyron relation (5.2) then yields

$$T_m - T_m^0 = \frac{\Delta V_m}{\Delta S_m}(P - P_0) = \frac{\Delta V_m}{T_m \Delta S_m} T_m^0 K_0 \frac{\Delta V}{V_0} \qquad (5.12)$$

The value of Kraut–Kennedy constant is therefore $C = K_0(\Delta V_m / L)$.

As pointed out by Gilvarry (1966), the Kraut–Kennedy relation is a linear approximation which, like the Simon equation, should properly be used as an interpolation formula only. Kraut and Kennedy (1966b) dispute this conclusion, and Kennedy and Vaidya (1970) indeed report that the linear Kraut–Kennedy relation gives a good fit for the experimental results on metals, but that the melting curves of van der Waals solids, concave toward the T-axis, are best fitted by the Simon law. The melting curves of ionic crystals and silicates are concave toward the $\Delta V / V_0$-axis. Use of the Kraut–Kennedy relation and a fortiori Simon relation for extrapolation would therefore overestimate the melting temperature of minerals.

5.4 Theoretical melting models

5.4.1 Shear instability models

Brillouin (1938, 1940) analyzed the thermal vibrations in strained solid bodies and gave a physical justification to the already old idea that, since liquids have no shear strength, the shear modulus of a solid should decrease with increasing temperature and vanish at the melting point (Sutherland 1891; Brillouin 1898). He remarked, however, that it would be difficult to imagine that the shear modulus should decrease smoothly up to the melting point, since this would imply melting without heat of fusion, that is, it would not account for the first-order character of melting.

Durand (1936) in the course of an experimental study of the elastic moduli of alkali halides and MgO had earlier observed that C_{12} was almost independent of temperature whereas C_{11} decreased linearly. He extrapolated the data to the temperature at which $C_{11} = C_{12}$, assumed to be the melting temperature (thus implicitly defining the melting point as the temperature for which the modulus $\mu = (C_{11} - C_{12})/2$ vanishes), and calculated values of the melting points quite close to the experimental values (e.g., 1170 K instead of 1077 K for NaCl).

Born (1939) unambiguously chose the loss of shear resistance as a criterion for melting and stated, "a theory of melting should consist of an investigation of the stability of a lattice under shearing stress." He calculated a second-order expansion of the free energy density of a cubic crystal as a function of finite strain and wrote the conditions for stability which guarantee that the energy is positive definite:

$$3K = C_{11} + 2C_{12} > 0; \qquad C_{11} - C_{12} > 0; \qquad C_{44} > 0 \qquad (5.13)$$

Born considered (without offering much justification for it) that melting occurs when the modulus C_{44}, corresponding to shear along the {100} planes, vanishes first, and that a gel would be produced if $C_{11} - C_{12}$, corresponding to shear along the {110} planes, vanished first. He then proceeded to calculate explicit expressions for the coefficients in the expansion of the free energy, using lattice dynamics within Debye's quasi-harmonic approximation, and determined the variation of the elastic moduli with T/Θ_D for different pressures and, finally, the melting curve T_m/Θ_D versus P. Born was quite aware that the latent heat of fusion could not be calculated from his theory by using Clapeyron's relation, since the change of volume on melting depended on the properties of the liquid state. It is also worth remarking that Born showed that his theory led to an expression of vibrational frequencies as a function of T_m similar to Lindemann's (see Sec. 5.4.2) with a slightly different numerical coefficient:

$$\nu_0 = CR^{1/2}N^{1/3}T_m^{1/2}M^{-1/2}V^{-1/3} \qquad (5.14)$$

May (1970), instead of plotting the shear moduli $C' = (C_{11} - C_{12})/2$, and C_{44} against temperature, plotted them for metals and argon, against $u = (V - V_0)/V_0$, where V and V_0 are the specific volume at temperature T and at 0°C, respectively. He found that the plots were straight lines and that C' vanished for a value of u slightly higher than but quite close to that of the melt (e.g., for Al, $C' = 0$ for $u = 0.125$, whereas melting occurs with $u = 0.123$).

Jackson and Liebermann (1974) calculated the critical temperatures for shear instability and their initial pressure dependences for alkali halides and oxides. They used the instability criterion $C' = 0$ for halides with

Table 5.3. *Comparison of the critical temperatures for shear instability T_C and melting point T_M and their initial pressure variation for oxides (OX) and their fluoride (FL) analogues*

Analogues	$\dfrac{T_C^{OX}}{T_C^{FL}}$	$\dfrac{T_M^{OX}}{T_M^{FL}}$	$\dfrac{dT_M^{FL}}{dP}$	$\dfrac{dT_C^{FL}}{dP}$	$\dfrac{dT_C^{OX}}{dP}$
LiF–MgO	2.8	2.8	12	10	11
NaF–CaO	2.4	2.3	15.1–18	14	13
KF–SrO	2.6	2.4	22.6	21	19

Source: From Jackson and Liebermann (1974).

NaCl structure ($B1$) and $C_{44} = 0$ for those with CsCl structure ($B2$). They found that, although the critical temperatures were higher than the melting temperatures by 30–200 K, the initial pressure dependence was the same as for melting; also, the critical temperatures and the initial pressure dependences are ordered in the same sense that the melting points and the initial slopes of the melting curves. There was an excellent correspondence between these parameters for oxides and for their "weakened" fluoride analogues (Table 5.3).

Tallon, Robinson, and Smedley (1977) and Tallon and Robinson (1977) modified the Born shear instability model to account for the first-order character of melting. They plotted the logarithm of the isothermal bulk modulus against the true dilatation $\ln V/V_0$, for solid alkali halides and for their melts. They found that the plot is linear over a large temperature range including the melting point and that one can write

$$\frac{\partial \ln K_T}{\partial \ln V} = -g_K$$

where g_K is analogous to a Grüneisen parameter. The main point is that the variation of the bulk modulus with dilatation is continuous through the melting point. From similar plots for C_{44} and C' (Fig. 5.8), extrapolations show that, for $B1$ halides, C_{44} is finite at the melting point and that C' vanishes at the melting point on the melt side. Using these considerations, Tallon et al. propose that the free energy of the system is a minimum when the elastic energy increase due to expansion is balanced by the entropic term $2RT$, where $2R$ is the "communal" entropy, corresponding

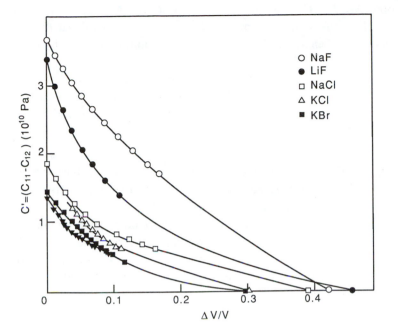

Figure 5.8. Evolution of shear modulus with dilatation for alkali halides. From top to bottom: NaF, LiF, NaCl, KCl, KBr (after Tallon et al. 1977).

to the fact that, when the shear modulus vanishes, the onset of fluidity allows the ions to have access to every part of the volume.

5.4.2 Vibrational instability: Lindemann law

i. Sutherland relation

At the end of last century, the idea of thermal vibrations of atoms in solids was well accepted. Sutherland (1890) thought that "at some characteristic temperature each solid ought to have a period of vibration characteristic of its molecule. The question is, at what temperature? At the melting point in each case the vibratory motion just breaks down, so that we ought to expect some simple relation amongst the periods of vibration of the elements at their melting points." He then proceeded to calculate the vibrational period at the melting point for an element of molecular mass M and specific heat C by writing that the kinetic energy of the molecule $\frac{1}{2}Mv^2$ is proportional to the heat received from 0 K to the melting point, MCT_m. The period was taken equal to the maximum amplitude possible at the melting point, $\alpha T_m V^{1/3}$ (where α is the linear thermal expansion coefficient and V is the molecular volume), divided by the mean velocity v.

Assuming that $MC = $ const. for elements and using an empirical relation $\alpha T_m V^{1/6} = $ const., which he claimed to have verified for metals, Sutherland obtained an expression for the vibrational period p:

$$p \propto V^{1/3} M^{1/3} T_m^{-1/2} \tag{5.15}$$

Using the data available in the literature, he concluded that "the periods of vibration of the molecules of solids at their melting points show very simple harmonic relations."

We have devoted some attention to Sutherland's paper to make it clear that, despite some recent contentions (e.g., Mulargia and Quareni 1988), Sutherland did not give a theory of melting and that the semi-empirical (and incorrect) relation (5.15) is not an earlier form of Lindemann's law, which we will now examine.

ii. Lindemann law

As a matter of fact, Lindemann (1910) did not propose a theory of melting either; his purpose was to calculate the Einstein vibrational frequency of a solid. He assumed a harmonic solid, whose atoms vibrate sinusoidally with a frequency

$$\omega = \left(\frac{\kappa}{m}\right)^{1/2} \tag{5.16}$$

where κ is the force constant and m is the mass of the atom.

As a step in his calculations, Lindemann assumes that, at the melting point, the amplitude of the vibration is so large that the atomic spheres collide, that is, the parameter δ, expressing the ratio of the distance between the surfaces of the spheres to the distance of their centers, vanishes. It is then possible to write that the integral of the work done on one atom by the restoring force κx (where x is the elongation), from the equilibrium position to the value $x = \delta r/2$, corresponding to the contact of two neighboring spheres of radius r, is equal to the thermal energy calculated with Einstein's model (Sec. 3.3.1):

$$\frac{\kappa \delta^2 r^2}{8} = \hbar\omega \left[\exp\left(\frac{\hbar\omega}{k_B T_m}\right) - 1\right]^{-1} \cong T_m - \frac{\hbar\omega}{2k_B} \tag{5.17}$$

Replacing κ by its expression as a function of ω from (5.16), solving for ω, and neglecting small terms, Lindemann finds

$$\omega = 2^{3/2}R^{1/2}\delta^{-1}\left(\frac{T_m}{Mr^2}\right)^{1/2} \tag{5.18}$$

The ratio δ is calculated from the Clausius–Mossotti equation, relating the dielectric constant to the number of atoms per unit volume, and it is assumed to be constant; hence, in cgs units,

$$\omega = 4.12 \times 10^{12}\pi T_m^{1/2}M^{-1/2}V^{-1/3} \tag{5.19}$$

where V is the atomic (molar) volume and $M = Nm$ is the atomic (molar) mass.

Lindemann used the melting points to calculate the frequencies, and at no point in the original paper did he suggest using relation (5.19) to predict melting points.

Gilvarry (1956a, 1957b) gave a firmer basis to the Lindemann equation. Instead of assuming that melting occurs when neighboring spheres collide, he wrote that the root mean square amplitude of atomic vibrations at fusion is a critical fraction f of the distance r_m of separation of nearest neighbor atoms:

$$\langle u^2 \rangle = f^2 r_m^2 \tag{5.20}$$

(Strangely enough, the introduction of the critical ratio f, without which Lindemann's law would not have been so successful, is usually attributed to Lindemann, although it is nowhere to be found in the original 1910 paper and clearly belongs to Gilvarry.)

In Debye's approximation, at high temperatures, for a monatomic solid of atomic mass M, we have (see Ziman 1965):

$$\langle u^2 \rangle \cong \frac{9k_BT}{M\omega_D^2} = \frac{9\hbar^2T}{Mk_B\Theta_D^2} \tag{5.21}$$

Hence, with (5.20),

$$T_m = f^2 \frac{k_B}{9\hbar^2}M\Theta_D^2 r_m^2 \tag{5.22}$$

which can be written in terms of the atomic volume V, assuming that $r_m = (V/N)^{1/3}$, as

▶ $$T_m = 0.00321f^2MV^{2/3}\Theta_D^2 \tag{5.23}$$

Noting that Debye's frequency can be expressed as a function of the elastic constants (see Sec. 3.3.1), Gilvarry (1956a) obtains an interesting expression for Lindemann's law:

$$RT_m = \Omega(\nu)KV \qquad\qquad (5.24)$$

where $\Omega(\nu)$ is a function of Poisson's ratio at the melting point, proportional to f^2, and K and V are the bulk modulus and the atomic volume at the melting point. The critical ratio f is calculated for a few metals from experimental data and found to be identical and equal to about 0.08, which supports the theory.

Ross (1969) reformulated the Lindemann law in terms of the statistical mechanical partition function of a Lennard-Jones–Devonshire cell model of a solid and found again that the Simon and Kraut–Kennedy relations could be derived from it.

Wolf and Jeanloz (1984) noted that in its usual form, the Lindemann law is derived using a quasi-harmonic Debye approximation valid only for monatomic solids. They found that the anharmonic contribution to the mean square atomic displacement can be quite large at the melting point and they gave a lattice dynamics formulation of the Lindemann criterion for polyatomic, anharmonic crystals; in many cases, however, due to the lack of experimental data on the frequency spectra, the complete formulation cannot be used.

Stacey and Irvine (1977a), starting from the definition of the thermodynamic gamma in Mie–Grüneisen's equation of state, equated the thermal pressure with the pressure increase that would obtain if melting occurred at constant volume

$$2\gamma\rho T_m \Delta S_m = K\rho\Delta V_m$$

and, with Clapeyron's equation, they obtained the relation

$$\frac{d\ln T_m}{dP} = \frac{2\gamma}{K}$$

which resembles equation (5.27) and, in these authors' view, gives a sound thermodynamic basis to Lindemann's law.

Despite such attempts, the single-phased approach to melting, derived from the Lindemann relation, has generally been criticized – and rightly so – on thermodynamic grounds: Melting is defined as a vibrational instability of the solid and no account is taken of the liquid phase (the same criticism can be leveled at the shear instability models). The only valid thermodynamic criterion of melting should be the equality of the free energies of the solid and liquid phases. However, in the absence of any good theory of melting, Lindemann's law represents a valuable tool. In fact, in the rather common case where the structure and coordination of the liquid is close to that of the solid and the free energies of both phases differ very little, one may find good justification for using Lindemann's law.

Martin and O'Connor (1977) experimentally determined the value of $\langle u^2 \rangle$ near the melting point for Al, Cu, and alkali halides (LiF, NaCl, KCl, KBr) by measuring the reduction in intensity of the elastically scattered component of a Bragg diffraction peak caused by thermal vibrations. They found that the Gilvarry critical ratio f is equal to about 0.08 for metals (in agreement with the value determined by Gilvarry (1956a)) and to about 0.11 for alkali halides. They concluded that the vibrational melting relationship can be applied to simple crystals of similar type but that the factor f varies with the crystal structure and the nature of the interaction force law.

Stern and Zhang (1988) determined $\langle u^2 \rangle$ in Pb using X-ray absorption spectroscopy (EXAFS) with synchrotron radiation. They found a value of the critical ratio at melting of $f = 0.068$. They also showed that in solid solutions of Hg in Pb, the Hg impurity atoms have a larger vibrational amplitude than the Pb atoms, and suggested that local premelting of the Hg–Pb bonds occurs below the bulk melting temperature (this may provide a physical reason for the lowering of melting point by impurities).

The case of polyatomic complex minerals is more difficult since the Lindemann–Gilvarry approach is properly restricted to Debye monatomic solids. It is, however, possible to use Lindemann's law (5.23) as a semi-empirical relation among melting temperature, acoustic Debye temperature, molar volume, and mean atomic mass $\bar{M}$. Poirier (1989) found that for 15 crystals with perovskite structure (fluorides and oxides) whose melting points and elastic moduli are known, there exists a good systematic correlation (Fig. 5.9) between T_m and $\bar{M}V^{2/3}\Theta_D^2$, where Θ_D is calculated using (3.37). For these crystals, the Gilvarry ratio is found to be $f = 0.11$ for perovskite oxides and about $f = 0.13$ for perovskite fluorides (Table 5.4).

Lindemann's law is, in most cases, used together with an equation of state (Gilvarry 1956c) to extrapolate melting points to high pressures, that is, obtain an estimate of the melting curve $T_m(P)$ or $T_m(V)$ when none is experimentally available. Wolf and Jeanloz (1984) extrapolated the experimental melting data for a few minerals to the (V, T) plane, using available thermodynamic data; they used Mie–Grüneisen's equation of state, taking into account the anharmonicity, and they found that the agreement between the predicted and experimental melting curves was poor or bad for most minerals (e.g., fayalite, diopside, pyrope), with the

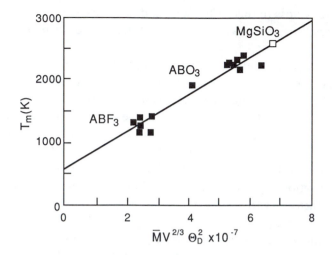

Figure 5.9. Correlation between actual melting temperature and the temperature predicted by the Lindemann law (proportional to the quantity plotted as abscissa) for oxide and fluoride perovskites: $ScAlO_3$, $GdAlO_3$, $SmAlO_3$, $EuAlO_3$, $YAlO_3$, $CaTiO_3$, $BaTiO_3$ (point in the middle), $SrTiO_3$, $KMgF_3$, $KMnF_3$, $KZnF_3$, $KNiF_3$, $KCoF_3$, $RbMnF_3$, and $RbCoF_3$ (after Poirier 1989).

exception of forsterite, for which the agreement was excellent. Mulargia and Quareni (1988) avoided using an equation of state by directly calculating T_m as a function of pressure from the values of the sound velocities at high pressures derived from experimental data. They compared the predicted slopes of the melting curves with the experimental ones for five metals, three alkali halides, and two minerals (bronzite and "peridotite" (?!)); they found that the agreement was satisfactory and concluded that the Lindemann law could profitably be used as a semi-empirical scaling law.

Another interesting way of using Lindemann's law to extrapolate melting curves consists in starting from its differential form giving the slope of the melting curve in terms of the Grüneisen parameter.

Gilvarry (1956a) showed that at high temperature and in the limit of Slater's approximation (Poisson's ratio constant along the melting curve), differentiation of equation (5.24) gives

$$\frac{d \ln T_m}{d \ln V} = \left(\frac{d \ln K}{d \ln V} \right)_{T = Tm} + 1$$

and with Slater's definition of gamma (at the melting temperature),

Table 5.4. *Critical Gilvarry–Lindemann factor f for various fluorides and oxides with perovskite structure*

ABX_3	$\bar{M}$	ρ	V_m	T_m	Θ_D	f
$KMgF_3$	24.08	3.15	4.29	1413	320	0.13
$KMnF_3$	30.21	3.42	3.43	1308	244	0.14
$KZnF_3$	32.26	4.02	3.43	1143	252	0.12
$KNiF_3$	30.99	3.99	3.53	1403	262	0.13
$KCoF_3$	31.00	3.82	3.38	1305	247	0.14
$RbMnF_3$	39.48	4.32	3.13	1259	221	0.13
$RbCoF_3$	40.31	4.76	3.31	1148	239	0.11
$ScAlO_3$	24.00	4.28	6.19	2143	513	0.11
$GdAlO_3$	46.45	7.44	4.43	2303	354	0.11
$SmAlO_3$	45.07	7.18	4.57	2373	364	0.11
$EuAlO_3$	45.39	7.25	4.41	2213	353	0.11
$YAlO_3$	32.78	5.35	5.09	2223	410	0.11
$CaTiO_3$	27.20	4.04	5.63	2248	438	0.11
$BaTiO_3$	46.25	6.04	3.77	1898	280	0.12
$SrTiO_3$	27.70	5.12	5.29	2213	403	0.11

Note: $\bar{M}$ is the mean atomic mass in grams per atom, ρ is the specific mass in grams per cm^3, V_m is the average Debye acoustic velocity in km/s, T_m is the melting point in K, and Θ_D is the Debye temperature in K, calculated from (3.27).

$$\gamma_S = -\frac{1}{6} - \frac{1}{2}\frac{d\ln K}{d\ln V}$$

we have

$$\frac{d\ln T_m}{d\ln V} = -2\gamma_S + \frac{2}{3} \tag{5.25}$$

or

▶
$$\frac{d\ln T_m}{d\ln \rho} = 2\left(\gamma_S - \frac{1}{3}\right) \tag{5.26}$$

or

$$\frac{dT_m}{dP} = \frac{T_m}{K}2\left(\gamma_S - \frac{1}{3}\right) \tag{5.27}$$

The same relations can of course be obtained by taking the logarithmic derivative of (5.23) and using the definition $\gamma = -d\ln\Theta_D/d\ln V$.

Note that

$$\frac{d \ln T_m}{d \ln \rho} = \frac{d \ln \Phi}{d \ln \rho}$$

where $\Phi = K/\rho$ is the seismic parameter.

Equation (5.26) is a useful differential form of Lindemann's law, which is very convenient for extrapolating melting curves. Its integrated form, taking into account the variation of γ with density ($\gamma\rho = \gamma_0\rho_0$), is

$$T_m = T_m^0 \exp\left[2\gamma_0\left(1 - \frac{\rho_0}{\rho}\right) + \frac{2}{3}\ln\left(\frac{\rho_0}{\rho}\right)\right] \tag{5.28}$$

It can be used with a theoretical or an experimental EOS.

Note that by writing (5.25) as a finite difference equation (taking into account the fact that $\Delta V < 0$)

$$\frac{\Delta T_m}{T_m^0} = \frac{T_m - T_m^0}{T_m^0} = 2\left(\gamma - \frac{1}{3}\right)\frac{\Delta V}{V}$$

one obtains the Kraut–Kennedy relation:

$$T_m = T_m^0\left[1 + \left(2\gamma - \frac{1}{3}\right)\frac{\Delta V}{V}\right]$$

where the constant C is expressed in terms of gamma (Gilvarry 1966).

We clearly see here that the Kraut–Kennedy relation replaces the melting curve in the T, V plane by its tangent at the origin and generally cannot be used for extrapolations to very high pressures.

5.4.3 *Lennard-Jones and Devonshire model*

The two essential characteristics of the liquid state are that it has a vanishing shear modulus and no long-range order, and indeed it can be shown that the loss of rigidity is a direct consequence of the loss of long-range order (P. W. Anderson 1984). We have seen that the problem with the shear instability melting theories is that one defines melting by the vanishing of the shear modulus of the solid, without reference to the melt and that, consequently, these theories have no sound thermodynamic grounding. No such inconvenience arises if one, instead, defines melting by the transition from an ordered to a disordered state, which can be thermodynamically defined. This is what Lennard-Jones and Devonshire (1939a, b) did in constructing what is probably the most rigorous theory of melting of simple solids.

An atom of the solid or the liquid is regarded as vibrating in the available space of a small cell (the free volume, defined in Sec. 3.5). Lennard-

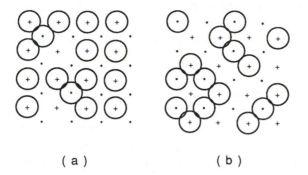

(a)   (b)

Figure 5.10. Melting by atomic disordering (after Lennard-Jones and Devonshire 1939). (a) Local displacement of atoms, but long distance order preserved. (b) Long distance order broken: equal number of atoms on regular lattice sites (crosses) and interstitial sites (dots).

Jones and Devonshire consider a crystal made up of identical atoms on regular lattice sites α; the interstitial sites β are positions of higher energy. The proportion of atoms on interstitial sites (Fig. 5.10) is taken as a measure of disorder: the order parameter Q is defined as the ratio of the number of atoms on α sites, N_α, over the total number of atoms N:

$$Q = \frac{N_\alpha}{N}$$

$$1 - Q = \frac{N_\beta}{N}$$

(5.29)

Q varies between $\frac{1}{2}$ for total disorder and 1 for perfect order.

In a first paper, Lennard-Jones and Devonshire (1939a) calculated the effect of disorder on the partition function of atoms by determining the probability of various configurations about atoms on α and β sites. They found a critical temperature at which long-range order disappears and they determined the additional free energy and pressure due to disorder, in terms of the extra energy W of a pair of atoms on neighboring α and β sites. To calculate the melting temperature, they expressed W in terms of interatomic potentials. This method gave good results for rare gases, but the calculations were rather elaborate and Lennard-Jones and Devonshire (1939b) proposed a simpler method derived from the Bragg–Williams theory for order–disorder transformations in binary metallic alloys. Simultaneously and independently, Frank (1939) published a paper developing exactly the same method, showing that a cooperative disorder of the Bragg–Williams type leads to the right kind of two-phase, first-order melting transition. However, he did not give detailed calculations, and we will now return to the Lennard-Jones and Devonshire (1939b) quantitative model.

Instead of calculating the configurations about a given atom, they assume that its environment is governed by the average state of order throughout the system. To change the position of an atom from an α-site to a β-site when all the atoms are on α-sites requires much more energy, owing to the repulsive field of the neighbors, than when the change is made simultaneously for several atoms. The phenomenon of interchange is therefore regarded as a cooperative one.

The partition function of N atoms in perfect order is equal to f^N, where f is the partition function of each atom of mass m vibrating in its cell, given by

$$f = \left(\frac{2\pi mkT}{h^2}\right)^{3/2} V_F \exp\left(\frac{-\Phi_0}{NkT}\right) \tag{5.30}$$

where V_F, the free volume, is given by (3.76) and Φ_0 is the potential energy of the system when all the atoms are in their equilibrium positions. The partition function of the disordered system is

$$Z = f^N \mathfrak{D}(Q) \tag{5.31}$$

where $\mathfrak{D}(Q)$ is a factor introduced to take account of the disorder and Q is the order parameter defined in (5.29).

Assuming that the disorder is homogeneous, an atom on a β-site is surrounded on average by zQ atoms on α-sites and an atom on an α-site is surrounded by $z(1-Q)$ atoms on β-sites, z being the number of β-sites adjacent to an α-site, equal to the number of α-sites adjacent to a β-site in the system considered by Lennard-Jones and Devonshire. Now, if the interaction energy of a pair of neighbors on α- and β-sites is W, the average extra energy of an atom due to its neighbors is $W_\alpha = Wz(1-Q)$ for an atom on an α-site and $W_\beta = WzQ$ for an atom on a β-site. Hence the energy required to transfer an atom from an α-site to a β-site is

$$\Delta W = W_\beta - W_\alpha = Wz(2Q-1) \tag{5.32}$$

The total energy of interaction due to the disorder is

$$N_\alpha W_\alpha = N_\beta W_\beta = zNWQ(1-Q) \tag{5.33}$$

Note that W is a function of the distance between two sites and therefore of the volume of the solid as a whole; this is how the cooperative feature of the model is introduced.

The number of configurations with N_α atoms on α-sites and N_β atoms on β-sites is

$$\gamma(Q) = \frac{N!}{(N-N_\alpha)!(N_\alpha!)} \times \frac{N!}{(N-N_\beta)!(N_\beta!)} = \frac{N!}{[NQ]! - [N(1-Q)]!}$$

The disorder factor $\mathfrak{D}(Q)$ is therefore

$$\mathfrak{D}(Q) = \gamma(Q) \exp\left[\frac{-zNWQ(1-Q)}{kT}\right] \tag{5.34}$$

and the partition function of the system of atoms vibrating in disordered cells is, from (5.31),

$$Z = f^N \gamma(Q) \exp\left[\frac{-zNWQ(1-Q)}{kT}\right] \tag{5.35}$$

The maxima of the partition function for given volume, temperature, and W are found for $dZ/dQ = 0$, which, if one uses Stirling's approximation $\ln(N!) \cong N \ln N - N$, can be written

$$\frac{zW(2Q-1)}{2kT} = \ln Q - \ln(1-Q) \tag{5.36}$$

and, if one remembers that $\operatorname{argtanh} x = \frac{1}{2}\ln[(1+x)/(1-x)]$, one can write (5.36) as

$$2Q - 1 = \tanh\left[\frac{zW(2Q-1)}{4kT}\right] \tag{5.37}$$

Equation (5.37) always has one root for $Q = \frac{1}{2}$ and, if $zW/4kT > 1$, it has another root $Q_{max} > \frac{1}{2}$, corresponding to a maximum of the partition function Z and given in terms of W/kT by equation (5.36). The Helmholtz free energy $F = -kT \ln Z_{max}$ is then

$$F = -NkT \ln f + zNWQ_{max}(1-Q_{max})$$
$$+ 2NkT[Q_{max} \ln Q_{max} + (1-Q_{max}) \ln(1-Q_{max})]$$

The free energy consists of two terms. The first one, $F' = -NkT \ln f$, corresponds to the perfectly ordered crystal and the second term,

$$F'' = zNWQ_{max}(1-Q_{max})$$
$$+ 2NkT[Q_{max} \ln Q_{max} + (1-Q_{max}) \ln(1-Q_{max})] \tag{5.38}$$

corresponds to the extra contribution of the disorder of the centers about which the atoms vibrate.

The internal energy, entropy, and pressure can be similarly separated into terms corresponding to the ordered crystal and extra terms due to the state of positional disorder.

Since $F'' = U'' - TS''$, the disorder contributions to the internal energy and entropy are

$$U'' = zNWQ_{max}(1 - Q_{max}) \tag{5.39}$$

$$S'' = -2[Q_{max} \ln Q_{max} + (1 - Q_{max}) \ln(1 - Q_{max})] \tag{5.40}$$

The extra pressure due to disorder $P'' = -(\partial F''/\partial V)_T$ is

$$P'' = -zNQ_{max}(1 - Q_{max})\left(\frac{dW}{dV}\right)_T - \left(\frac{\partial F''}{\partial Q}\right)_{V,T}\left(\frac{\partial Q}{\partial V}\right)_T \tag{5.41}$$

Since F'' is a maximum for Q_{max}, the last term vanishes and, as W is a function of volume alone, we have

$$P'' = -zNQ_{max}(1 - Q_{max})\frac{dW}{dV} \tag{5.42}$$

Denoting by $n'' = zNQ_{max}(1 - Q_{max})$ the number of atoms in adjacent α- and β-sites, we have

$$P'' = -n''\frac{dW}{dV}$$

Since Q_{max} is a function of W/kT only, as seen in equation (5.36), and W is a function of volume only, it follows that, for a given temperature, n'' is a function of volume only and that P'' rises from zero for small volumes to a maximum and decreases to zero again for large volumes.

Owing to the existence of the term P'' in the equation of state, the shape of the isotherm is drastically modified from that of the ordered solid (Fig. 5.11a). The curve of free energy F as a function of volume is shown in Figure 5.11b for solid argon. It displays a maximum B between two minima A and C. The minimum A, for small volume, corresponds to a relatively ordered state (solid) and the other minimum C corresponds to nearly total disorder (liquid). Between A and C the system will have less free energy by following the straight line AC corresponding to a two-phased mixture of solid and liquid in equilibrium. The volume of melting can be read directly as the difference of the abscissae of the two minima. Now, the total pressure depends on volume and temperature; if the interaction energy W varies with volume as $W = W_0(V_0/V)^m$ and if $\Phi_0 \propto W_0$, it can be shown that $P = f(V/V_0, kT/\Phi_0)$. Therefore, there exists a functional relationship between V/V_0 and kT/Φ_0 at $P = 0$.

As we have seen (Fig. 5.11), P goes to zero for three values of V/V_0, the smallest one corresponding to the volume of the solid at melting, and it

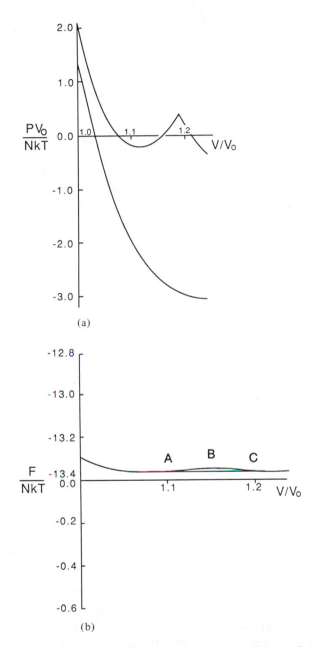

Figure 5.11. Lennard-Jones and Devonshire theory of melting (after Lennard-Jones and Devonshire 1939). (a) Pressure as a function of volume at a given temperature. The lower curve gives P', the pressure for a state of order, and the upper one gives the total pressure, the sum of P' and P'' the disorder pressure. (b) The free energy as a function of volume.

is found that for many solid gases the corresponding value of kT/Φ_0 is about 0.7. The melting temperature is therefore $T_m \cong 0.7\Phi_0/k$.

There is a good agreement between observed and calculated values of the temperature and volume of melting for solid gases.

Lennard-Jones and Devonshire (1939b) calculated the vibrational frequency of a solid with an interatomic potential (Lennard-Jones potential) given by

$$\Phi = \Phi_0[(r/r_0)^{-12} - 2(r/r_0)^{-6}] \qquad (5.43)$$

They were able to find a correlation between the vibrational frequency of the solid and the melting temperature calculated by their model, thus verifying Lindemann's law. They found that for a number of solid gases

$$T_m = \left(\frac{1}{163}\right)^2 \Theta_D^2 M V^{2/3} \qquad (5.44)$$

in good agreement with formula (5.23), if the Gilvarry factor f is taken equal to about 0.1.

5.4.4 Dislocation-mediated melting

The loss of long-range order is the fundamental feature of the melting phenomenon and, as we have seen in the case of the Frank and Lennard-Jones and Devonshire models, the cooperative appearance of disorder (the order parameter being a function of the total volume) accounts for the first-order character of the melting transition.

However, for energetic reasons, long-range order cannot be destroyed simultaneously over the whole crystal and disorder must be caused by the proliferation of lattice defects that locally break the order. This is the origin of defect-mediated theories of melting.

The simplest defect is the vacancy, a vacant lattice site. The equilibrium atomic fraction of vacancies in a crystal increases with temperature as $\exp(-\Delta G/kT)$, where ΔG is the free energy of formation of vacancies (see, e.g., Poirier 1985, p. 40). It is therefore tempting to envision melting as due to the increase in the number of vacancies up to a point where the vacancy-filled solid becomes so disordered that it can be considered as a liquid. However, the cooperative effects necessary to produce a first-order transition would appear only for unrealistically high vacancy concentrations.

Dislocations (see, e.g., Friedel 1964; Nabarro 1967; Sec. 6.3) are linear lattice defects that break the order along their cores. They also have a long-range stress field and cause elastic energy to be stored in the volume of the crystal. Motion of dislocations in the lattice causes slip and plastic deformation. It is easy to see how a crystal filled with dislocations can be so disordered and fluid as to be liquid-like. It probably was Shockley (1952) who first regarded the liquid as being a solid densely packed with dislocations and calculated its viscosity by consideration of the motion of dislocations.

Dislocation-melting theories indeed rest on the assumption that a liquid is a solid saturated with dislocation cores (Fig. 5.12) (Mizushima 1960; Kuhlmann-Wilsdorf 1965; Ninomiya 1978; Cotterill 1980; Walzer 1982; Suzuki 1983; Poirier 1986). The dislocation-melting models invoke a cooperative effect: The free energy of the solid increases with dislocation density but the introduction of new dislocations becomes easier and easier and, for a finite density, the free energy of the solid saturated with dislocations (i.e., a liquid) is equal to that of the dislocation-free solid. A first-order transition is then thermodynamically possible between solid and liquid at equilibrium.

The various dislocation-melting theories are reviewed by Poirier (1986), and a model based on Ninomiya's (1978) model is developed, giving the melting volume, entropy, and temperature as a function of the elastic moduli and the Grüneisen parameter. We will give here only the main outline of the model.

The total elastic internal energy ΔE of one mole of crystal containing a dislocation concentration C_d consists of:

The core energy

$$\Delta E_c \propto C_d b^3 \tag{5.45}$$

C_d is defined by $C_d = LV/Nb$, where L is the dislocation length per unit volume, V is the molar volume, N is Avogadro's number, and b is the Burgers vector of the dislocations (a vector of the Bravais lattice, expressing the strength of the dislocation).

The elastic energy stored in the bulk of the crystal, which can be written, taking into account the interaction of dislocations (Poirier 1986),

$$\Delta E_e \propto -C_d \mu b^3 \ln\left(\frac{\pi r_0^2 N C_d b}{V}\right)^{1/2} \tag{5.46}$$

where μ is the shear modulus and r_0 is the core radius.

Now the strain field of a dislocation has a dilatation component and the concentration C_d of dislocations introduces a dilatation $\epsilon(C_d)$, which, in turn, causes the shear modulus to decrease. Using Slater's formula (3.66), we can write

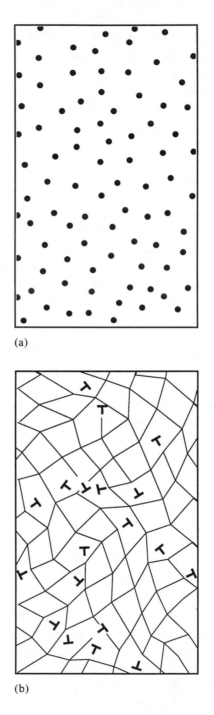

(a)

(b)

Figure 5.12. Dislocation melting (two-dimensional example): a disordered liquid (a) can be represented by a solid saturated with dislocations (b); the dislocation cores are traditionally represented by "nails."

$$\mu = \mu_0[1 + (2\gamma + \tfrac{1}{3})\epsilon] \tag{5.47}$$

where γ is Slater's gamma, and carry the value of μ into (5.46). A third energy term corresponds to the dilatation ϵ:

$$\Delta E_\epsilon \propto K\epsilon^2 \tag{5.48}$$

where K is the bulk modulus.

The total energy is therefore

$$\Delta E = \Delta E_c + \Delta E_e + \Delta E_\epsilon \tag{5.49}$$

The equilibrium value of ϵ corresponding to C_d is obtained by minimizing ΔE with respect to ϵ. For $C_d = C_d^{\text{sat}} \cong \tfrac{1}{3}$ at saturation, we have

$$\epsilon^{\text{sat}} \cong \frac{A}{2\pi} \frac{\mu_0}{K} \left(\gamma - \frac{1}{3}\right) C_d^{\text{sat}} \tag{5.50}$$

where A is a geometric factor depending on the crystal structure.

The melting volume is then

$$\Delta V_m = V\epsilon^{\text{sat}} \tag{5.51}$$

The melting entropy can be written

$$\Delta S = R[3\epsilon^{\text{sat}}\gamma + 2C_d^{\text{sat}}] \tag{5.52}$$

The first term corresponds to the entropy increase due to the lowering of lattice frequencies caused by anharmonic dilatation, and the second term corresponds to the vibrational entropy of dislocation lines.

The total extra free energy of the crystal with a dislocation concentration C_d, at temperature T, is

$$\Delta F = \Delta E - T\Delta S \tag{5.53}$$

and the melting temperature is obtained by taking $C_d = C_d^{\text{sat}}$ in (5.53) and setting $\Delta F = 0$ (Fig. 5.13), since ΔF corresponds to the difference in free energy between the liquid (dislocation saturated crystal) and the dislocation-free solid. We have

$$T_m = \frac{KV}{2R} \epsilon^{\text{sat}} \frac{1 - \epsilon^{\text{sat}}(\gamma - \tfrac{1}{3})}{(\gamma - \tfrac{1}{3})(3\epsilon^{\text{sat}}\gamma + 2C_d^{\text{sat}})} \tag{5.54}$$

Using (5.51), (5.52), and the Clapeyron equation, we obtain the slope of the melting curve

$$\frac{dT_m}{dP} = \frac{\Delta V_m}{\Delta S_m} = \frac{V}{R} \epsilon^{\text{sat}} (3\epsilon^{\text{sat}}\gamma + 2C_d^{\text{sat}})^{-1} \tag{5.55}$$

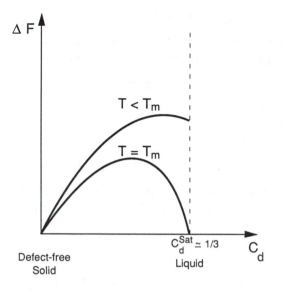

Figure 5.13. Extra free energy of a crystal containing a dislocation concentration C_d. At the melting temperature, $\Delta F = 0$ for $C_d = C_d^{\text{sat}}$.

and

$$\frac{d \ln T_m}{d \ln \rho} = \frac{2(\gamma - \frac{1}{3})}{1 + \epsilon^{\text{sat}}(\gamma - \frac{1}{3})} \tag{5.56}$$

We see that equation (5.56) differs from the differential Lindemann law (5.26) only by the coefficient $1/[1 - \epsilon^{\text{sat}}(\gamma - \frac{1}{3})]$, which is slightly larger than 1 if $\gamma > \frac{1}{3}$, which is always the case.

The melting parameters of iron were calculated using this model (Poirier 1986) and were found to be in good agreement with the experimental values, as we will see in next section.

Stacey and Irvine (1977a, b) derive the differential Lindemann law from what they claim to be a dislocation theory of melting and Clapeyron's relation. However, their model has no cooperative feature and the liquid state is not characterized by a saturation concentration of dislocations; instead, they assume that the latent heat of melting and the melting volume are respectively equal to the extra energy and anharmonic dilatation corresponding to the removal of an atom from one of a pair of linear mass and spring type chains – a system only very remotely connected to a crystal dislocation.

In conclusion, all melting theories are imperfect because to construct a good theory for the first-order melting transition, one must write that the free energies of the solid and the liquid are equal at equilibrium. This would necessitate a better description of the liquid state than the ones currently available – indeed, the structure of the liquid state is a standing problem in physics. Still, such as they are, the disorder theories of melting of the Frank–Lennard-Jones–Devonshire type or the dislocation theories of melting are the best.

The shear instability theories, even though they provide some interesting correlations, are not good theories of melting; nor are the vibrational theories in which, to quote Frank (1939), one assumes that "at the melting point, the crystal shakes itself to pieces." We must, however, remark that Lindemann (1910) never claimed to have done anything other than find a correlation between the melting point and the vibrational frequency, and it is no surprise that the Lindemann law can be found as a by-product of any theory of melting worth its salt, even though that does not constitute a justification of a "Lindemann theory of melting." As a correlation, and especially in its differential form, in terms of the Grüneisen parameter, Lindemann's law is very valuable and its status is not impaired by the fact that it does not embody a theory of melting.

It remains that even the best theories of melting, developed for simple solids such as solid gases or metals, dismally fail in the case of minerals due to the complexity of the structure of their liquids. Lindemann correlation, however, if applied to crystals of the same structure, may be the basis of fruitful systematics and reasonable extrapolations of the melting curve.

5.5 Phase diagram and melting of iron

There is strong evidence for the Earth's core being essentially composed of iron (Birch 1952; O. Anderson 1985), and, as we will see in what follows, the knowledge of the melting curve of iron at core pressures can constrain the temperature at the inner core boundary and anchor the Earth's temperature profile. The melting curve, however, is part of the phase diagram (see Sec. 7.4) and represents the equilibrium boundary between the liquid and the solid phase stable at a given pressure. It is therefore the whole phase diagram and not just the melting curve at ambient pressure that must be extrapolated at high pressure.

At pressures lower than about 200 kbar, the phase diagram of iron is reasonably well known (Guillermet and Gustafson 1984; Boehler 1986) (Fig. 5.14): The phase stable at room temperature and ambient pressure is α-Fe with a body-centered (bcc) structure. At high temperature, it changes into face-centered cubic (fcc) γ-Fe, which reverts to a bcc δ-Fe phase

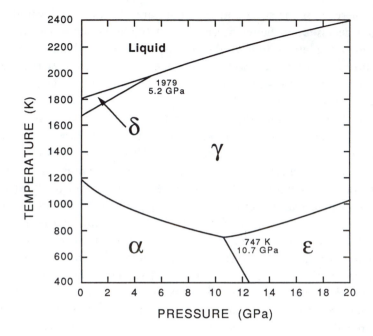

Figure 5.14. Phase diagram of iron (after Guillermet and Gustafson 1984).

below the melting point (Strong, Tuft, and Hanneman 1973). At high pressure, α-Fe transforms into hexagonal-close-packed (hcp) ϵ-Fe. The α to ϵ transformation has been studied in situ by X-ray diffraction, using synchrotron radiation, and appears to take place by a diffusionless mechanism (Bassett and Huang 1987). The experimental values of the triple point between α, γ, and ϵ phases are given in Table 5.5. At low and moderate pressures, the phase that melts is γ-Fe, and the melting curve has been determined up to 200 kbar (Liu and Bassett 1975) and 430 kbar (Boehler 1986) by resistive heating of an iron wire in a diamond–anvil cell and up to 1 Mbar (Williams et al. 1987) by laser heating of iron in a diamond–anvil cell. Shock-wave experiments (Brown and McQueen 1980, 1982, 1986; Williams et al. 1987) allow the determination of the melting point at one pressure only, about 2.5 Mbar.

Because the slope of the ϵ–γ boundary is higher than the slope of the melting curve of γ-Fe, there must be a triple point between ϵ, γ, and liquid. The pressure and temperature of the triple point can, however, be estimated only by extrapolations. If the triple point occurs at a pressure lower than the inner-core boundary pressure (3.3 Mbar), the inner-core

Table 5.5. *Pressure and temperature of the triple points*
of the phase diagram of iron

Reference	α–γ–ϵ		ϵ–γ–liquid	
Bundy (1965)	110 kbar	763 K		
Liu (1975b)			940 kbar	3243 K
Anderson (1986)			2800 kbar	5760 K
Boehler (1986)	116 kbar	810 K	750 kbar	2773 K
Akimoto et al. (1987)	83 kbar	713 K		

boundary temperature corresponds to the melting of ϵ-Fe, whereas it corresponds to the melting of γ-Fe if the pressure of the triple point is higher than 3.3 Mbar (Birch 1952; O. Anderson 1986).

There still is no consensus on the pressure of the ϵ–γ–liquid triple point (Table 5.5): It is thought to be higher than 3.3 Mbar by O. Anderson (1982), and lower than 3.3 Mbar by Liu (1975b), Brown and McQueen (1982), O. Anderson (1986), and Mao, Bell, and Hadidiacos (1987). From his experimental data, Boehler (1986) expects it to be at 750 kbar and 2500 K, in which case there could be a new iron phase at pressures higher than 2 Mbar and it would be the phase of the inner core.

Predicted temperatures of the melting point of pure iron at the pressure of the inner core boundary obtained theoretically or through extrapolations of experimental data, using Kraut–Kennedy, Lindemann, or other laws, are spread over a large temperature interval, although most of them tend to crowd about 6000 K (see O. Anderson 1986; Mulargia 1986; Poirier 1986) with the exception of the large value (7000 K) obtained by extrapolation of experimental data by Williams et al. (1987) (Table 5.6, p. 132).

Impurities in solution usually lower the melting temperature of iron (see Sec. 5.1). This is, in particular, the case for sulfur (Fig. 5.2) and silicon. For a sulfur content of 8 wt%, the depression of the melting point at the pressure of the inner core boundary is about 1100 K (Stevenson 1981).

Table 5.6. *Melting point of pure iron at the pressure of inner core boundary (3.3 Mbar)*

Reference	T_m (K)
Gilvarry (1957b)	6200
Higgins and Kennedy (1971)	4250
Liu (1975b)	4920
Boschi (1974)	6600
Bukowinski (1977)	5450
Stevenson (1981)	6300
Brown and McQueen (1986)	5800
Spiliopoulos and Stacey (1984)	6140
Anderson (1986)	6210
Poirier (1986)	6150
Williams, Jeanloz, et al. (1987)	7600

6

Transport properties

6.1 Generalities

All dynamic processes inside the Earth are governed by the transport of certain physical quantities, or at least depend on material constants (conductivities or diffusivities) that express how easily those quantities are transported in a given region of the Earth.

Solid-state reactions between minerals and kinetics of phase transformations are controlled by transport of matter (atoms) by diffusion.
Viscous flow of matter in the convecting mantle is controlled by diffusion of momentum (viscosity).
Cooling of the Earth is controlled by heat transfer in the thermal boundary layers.
The propagation of electromagnetic signals in the mantle depends on its electrical conductivity.

Transport of a physical quantity (e.g., momentum or heat) or of particles (e.g., atoms or electrically charged particles) always results from the application of a *driving force* $\mathbf{F}$. In all cases it produces a *flux* $\mathbf{J}$ (measured per unit area and unit time).

In the case of particles, the velocity $\mathbf{v}$ of the particle depends on the driving force and in the linear approximation, we have

$$\mathbf{v} = M\mathbf{F} \qquad (6.1)$$

where M is the *mobility* of the particle, which depends on the nature of the particle, on the properties of the medium in which it moves, and on temperature and pressure. The flux $\mathbf{J}$ is then equal to the number of particles per unit volume times the velocity.

In general, if the driving force derives from a potential U, $\mathbf{F} = -\operatorname{grad} U$, the linear approximation gives

133

$$\mathbf{J} = -C \operatorname{grad} U \qquad (6.2)$$

where C is a material constant expressing the ease of transport.

The driving force may be external and imposed on the system (e.g., flow due to a pressure gradient, electric current due to a difference of potential). It can also reflect the existence of heterogeneities in the system, which the resulting flow will tend to eliminate (e.g., diffusion of matter, diffusion of momentum or viscosity, diffusion of heat). In that case, as we will see, equations of the type (6.2) together with conservation equations yield diffusion equations.

Diffusion is an out-of-equilibrium, dissipative, irreversible process. Steady-state flow in such processes corresponds to a minimum of the rate of entropy production dS/dt. It can be shown (Prigogine 1962) that the rate of entropy production can be written as the product of a generalized driving force $\mathbf{F}$ by the corresponding flux $\mathbf{J}$.

Let us now consider a few important cases of transport phenomena mentioned above:

i. Fluid flow through a porous medium

The external driving force is a pressure gradient. The flow of an incompressible fluid per unit area and per second is given by *Darcy's equation*

$$\mathbf{J} = -\frac{k}{\eta} \operatorname{grad} P \qquad (6.3)$$

where k is the permeability of the medium, expressed in Darcys (1 Darcy = $1\mu m^2$) and η is the dynamical viscosity.

ii. Electrical conduction

The external force is an electric field deriving from a potential, $E = -\operatorname{grad} V$. The transport equation is *Ohm's law*

$$\mathbf{J} = \sigma_{el} E = -\sigma_{el} \operatorname{grad} V \qquad (6.4)$$

where σ_{el} is the electrical conductivity, expressed in Siemens/m (1 S/m = $1\ \Omega^{-1}m^{-1}$).

The electric current density $\mathbf{J}$ is equal to the number of charge-carrying particles per unit volume times the electric charge they carry times their velocity.

iii. Diffusion of matter

The driving force for the diffusion of species A (atom or point defect) is the gradient of its chemical potential μ_A. The flux of matter is given by:

$$\mathbf{J}_A = -L_A \operatorname{grad} \mu_A \tag{6.5}$$

where $\mathbf{J}_A$ is the flux of atoms A per second, L_A is called the phenomenological coefficient. It is, however, more convenient to use *Fick's equation,* expressed in terms of the concentration c_A of atoms (number of atoms per unit volume),

$$\blacktriangleright \qquad \mathbf{J}_A = -D_A \operatorname{grad} c_A \tag{6.6}$$

where D_A is the *diffusivity* or diffusion coefficient of species A, expressed in m²/s or cm²/s.

Concentration c_A is defined as the number of atoms of species A per unit volume. If n is the total number of atoms per unit volume in the medium and N_A is the atomic fraction of atoms A, we have

$$nN_A = c_A \tag{6.7}$$

By definition, the chemical potential is

$$\mu_A = \mu_A^0 + RT \ln \gamma_A N_A \tag{6.8}$$

where μ_A^0 is the chemical potential in the standard state and γ_A is the activity coefficient of species A. In the simple case of ideal solutions, $\gamma_A = 1$ and we have

$$\operatorname{grad} \mu_A = \frac{\partial \mu_A}{\partial x} = \frac{\partial \mu_A}{\partial N_A} \frac{\partial N_A}{\partial x} = \frac{RT}{nN_A} \frac{\partial c_A}{\partial x} \tag{6.9}$$

Hence

$$\mathbf{J}_A = -L_A \operatorname{grad} \mu_A = -L_A \left(\frac{RT}{c_A} \right) \operatorname{grad} c_A \tag{6.10}$$

and, by comparison with (6.6),

$$D_A = \frac{L_A RT}{c_A} \tag{6.11}$$

Now, from (6.1), we have the velocity $\mathbf{v}_A = -M_A \operatorname{grad} \mu_A$ and, since $\mathbf{J} = c_A \mathbf{v}_A$, we have $L_A = c_A M_A$. Substituting into (6.11), we obtain *Einstein's relation,* linking the diffusivity with the mobility,

$$D_A = RTM_A \tag{6.12}$$

The conservation equation for species A is written

$$\operatorname{div} \mathbf{J}_A + \frac{\partial c_A}{\partial t} = 0 \tag{6.13}$$

Hence, with equation (6.6),

$$\frac{\partial c_A}{\partial t} = \operatorname{div}(D_A \operatorname{grad} c_A) \tag{6.14}$$

If diffusivity is independent of concentration (an approximation generally valid only for dilute systems), we obtain the diffusion equation (2nd Fick's equation)

▶
$$\frac{\partial c_A}{\partial t} = D_A \nabla^2 c_A \tag{6.15}$$

where $\nabla^2 = \operatorname{div} \operatorname{grad}$ is the Laplacian operator in cartesian coordinates.

The irreversibility of diffusion is clearly seen in the equation, since it is not invariant under transformation of t into $-t$.

iv. Diffusion of momentum

The viscosity of a flowing fluid is due to the diffusional transfer of momentum per unit volume ρv (ρ is the specific mass and v is the velocity) down the momentum gradient, that is, from regions of higher velocity to regions of lower velocity in an incompressible fluid.

The linear relation (6.1) can be written here as

$$\sigma = -\nu \operatorname{grad}(\rho v) \tag{6.16}$$

σ is the shear stress (force per unit area) equal to the flux of momentum (momentum transferred per unit area and per second). ν is the *kinematic viscosity* expressed in m^2/s or cm^2/s.

Equation (6.16) can be written:

$$\sigma = \eta \dot{\epsilon} \tag{6.17}$$

where $\eta = \rho v$ is the *dynamic viscosity* expressed in Pa s or Poises (1 Pa s = 10 Poise) and $\dot{\epsilon} = d\epsilon/dt = -\operatorname{grad} v$ is the shear rate (Fig. 6.1). This is *Newton's relation*, which often serves as a definition of the viscosity.

The corresponding diffusion equation of momentum is a simple form of the Navier–Stokes equation for an incompressible fluid:

$$\frac{\partial(\rho v)}{\partial t} = \nu \nabla^2(\rho v) \tag{6.18}$$

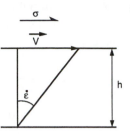

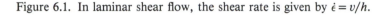

Figure 6.1. In laminar shear flow, the shear rate is given by $\dot{\varepsilon} = v/h$.

v. Diffusion of heat

The driving force here is (see Callen 1985)

$$\mathbf{F} = -\mathrm{grad}\left(\frac{\partial S}{\partial Q}\right) = -\mathrm{grad}\left(\frac{1}{T}\right) \qquad (6.19)$$

where S is the entropy, T is the temperature, and Q is the internal energy (heat). The linear equation (6.2) can therefore be written:

$$\mathbf{J} = kT^2\,\mathrm{grad}\left(\frac{1}{T}\right)$$

or

$$\mathbf{J} = -k\,\mathrm{grad}\,T \qquad (6.20)$$

where k is the thermal conductivity. The diffusion equation is *Fourier's equation*

$$\frac{\partial T}{\partial t} = \frac{k}{\rho C_p}\nabla^2 T \qquad (6.21)$$

where C_p is the specific heat at constant pressure. It can also be written

$$\frac{\partial T}{\partial t} = \kappa\nabla^2 T \qquad (6.22)$$

where $\kappa = k/\rho C_p$ is the *thermal diffusivity*, expressed in m²/s or cm²/s.

The solutions of the differential equation for the diffusion of heat have been determined for a great variety of initial and boundary conditions by Carslaw and Jaeger (1959) and can of course be applied to the diffusion of matter.

As an example, let us give here the solution of Fourier's equation in the case of a semi-infinite medium bounded by the plane $x = 0$ (extending to infinity in the sense of $x > 0$), whose initial temperature at time $t = 0$ is a constant T_0, the surface ($x = 0$) being kept at $T = 0$. The temperature T of the medium at time t and depth x is given by

$$T = T_0 \, \text{erf} \left[\frac{x}{2(\kappa t)^{1/2}} \right] \tag{6.23}$$

where erf is the (tabulated) error function defined by:

$$\text{erf}(y) = \frac{2}{\sqrt{\pi}} \int_0^y \exp(-u^2) \, du \tag{6.24}$$

We see that since the temperature at a given time and depth depends only on the dimensionless parameter $x/(\kappa t)^{1/2}$, the time required for a point to reach a given temperature is proportional to the square of its distance to the surface, for example, the time to reach $T = \frac{1}{2}T_0$ is

$$t = \frac{1.099}{\kappa} x^2 \tag{6.25}$$

Incidentally, these are the simplified boundary conditions Kelvin used to estimate the age of the Earth (see Carslaw and Jaeger 1959). The temperature gradient, from (6.23) and (6.24), is

$$\frac{\partial T}{\partial x} = T_0 (\pi \kappa t)^{-1/2} \exp \left[\frac{-x^2}{4\kappa t} \right] \tag{6.26}$$

Hence the geothermal gradient at the surface is

$$G = T_0 (\pi \kappa t)^{-1/2} \tag{6.27}$$

Taking $T_0 = 1200°C$ (for molten rock) and a typical value of $\kappa = 0.118$ cm^2/s, Kelvin found that it took about 9 million years for the geothermal gradient at the surface to reach its present value taken to be equal to be 37°C/km. Note, however, that besides not taking into account the production of heat by radioactive decay, which he could not know, Kelvin did not really calculate the time it would take the whole Earth to cool down, and that using the same numerical values and equation (6.25), we find that it would take more than 10^9 years to reduce by half the temperature at a depth of 200 km.

vi. Dimensionless numbers

The diffusion coefficient D, the kinematic viscosity ν, and the thermal diffusivity κ have the same dimensions and represent material properties. Dimensionless numbers characterizing a fluid can be constructed with these parameters (see, e.g., Tritton 1977). Particularly interesting are the *Prandtl number* $\mathcal{P} = \nu/\kappa$ and the *Schmidt number* $\mathcal{S} = \nu/D$.

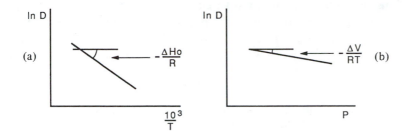

Figure 6.2. (a) Typical Arrhenius plot for diffusion. The activation energy ΔH is given by the slope of the straight line. (b) Typical variation of diffusion coefficient with pressure. The activation volume is given by the slope of the straight line.

6.2 Mechanisms of diffusion in solids

It is a fact of experience that diffusivity in solids varies with temperature according to an Arrhenius law (Fig. 6.2)

$$D = D_0 \exp\left(\frac{-\Delta H}{RT}\right) \qquad (6.28)$$

where ΔH is the activation enthalpy for diffusion, given by the slope of the Arrhenius plot:

$$\Delta H = -R \frac{\partial \ln D}{\partial(1/T)} \qquad (6.29)$$

The activation enthalpy is a function of pressure, and the pressure dependence of the diffusion coefficient is expressed by an apparent activation volume:

$$\Delta H = \Delta H_0 + P\Delta V \qquad (6.30)$$

$$\Delta V = -RT \frac{\partial \ln D}{\partial P} \qquad (6.31)$$

The linear expression (6.30) is of course valid only at relatively low pressures; at pressures of the lower mantle, the pressure dependence of the activation volume would have to be taken into account.

In most cases relevant to geophysics, transport of atoms is effected by exchange with vacant lattice sites or *vacancies,* which migrate through the crystal.

Let us consider the important case of *self-diffusion,* that is, diffusion of an atom A in a crystal constituted totally or partially of atoms A (e.g., diffusion of Fe in Fe or diffusion of O in MgO).

The probability for an atom to exchange with a vacancy, that is, to diffuse by one jump, is equal to the probability of finding a vacancy in a neighboring site multiplied by the probability of successfully jumping into it. Hence the self-diffusion coefficient is given by

$$D_{sd} = N_v D_v \tag{6.32}$$

where N_v is the equilibrium atomic fraction of vacancies at temperature T and D_v is the diffusion coefficient of vacancies.

It can be shown (see, e.g., Poirier 1985) that the equilibrium concentration of (thermal) vacancies is

$$N_v = \exp\left(\frac{-\Delta G_f}{RT}\right) \tag{6.33}$$

with

$$\Delta G_f = \Delta H_f - T\Delta S_f \tag{6.34}$$

where ΔG_f, ΔH_f, and ΔS_f are the Gibbs free energy of formation, the enthalpy of formation, and the entropy of formation of vacancies, respectively.

Vacancies diffuse by a random walk process described by another Einstein's relation (there are many Einstein's relations!):

$$\langle R^2 \rangle = \Gamma t (\delta l)^2 \tag{6.35}$$

where $\langle R^2 \rangle$ is the mean square distance covered by a vacancy during time t, Γ is the jump frequency, and δl is the jump distance. The diffusion coefficient of vacancies is defined by

$$D_v = \frac{\langle R^2 \rangle}{6t} = \frac{1}{6}\Gamma(\delta l)^2 \tag{6.36}$$

Note that the root mean square distance covered in time t by vacancies (or any particle diffusing by random walk) is proportional to the square root of time, as expected from (6.23),

$$x \propto (Dt)^{1/2} \tag{6.37}$$

The jump frequency Γ is thermally activated, with an activation enthalpy equal to the activation enthalpy for migration of vacancies ΔH_m. Hence

$$\Delta H_{sd} = \Delta H_f + \Delta H_m \tag{6.38}$$

In ionic compounds, each ionic species diffuses on its own sublattice and one can define the coefficients of self-diffusion of cations D_c and of anions D_a. In the case of binary compounds, we have

$$D_c = D_{cv} N_{cv} \tag{6.39}$$

$$D_a = D_{av} N_{av} \tag{6.40}$$

where D_{vc} and D_{av} are the diffusion coefficients of cationic and anionic vacancies, respectively.

Cationic vacancies are negatively charged and anionic vacancies are positively charged; they are formed in electrically neutral pairs (Schottky pairs). In pure ionic crystals, at equilibrium, the product of the atomic fractions of cationic and anionic vacancies depends only on temperature:

$$N_a N_c = N_0^2 = \exp\left(\frac{\Delta S_f}{R}\right)\exp\left(\frac{-\Delta H_f}{RT}\right) \tag{6.41}$$

with, for electrical neutrality,

$$N_a = N_c$$

The activation enthalpies for diffusion of cations and anions are then

$$\Delta H_c = \tfrac{1}{2}\Delta H_f + \Delta H_{mc} \tag{6.42}$$

$$\Delta H_a = \tfrac{1}{2}\Delta H_f + \Delta H_{ma} \tag{6.43}$$

Now, there can be, and often are, aliovalent cationic impurities in solution (e.g., Ca^{2+} replacing Na^+ ions in NaCl). If a Na^+ is replaced by a Ca^{2+}, there is an excess of positive charge that must be compensated for by the creation of one negatively charged cationic vacancy. If the atomic fraction of impurities is C, electrical neutrality demands that

$$N_a + C = N_c \tag{6.44}$$

Hence, with (6.41),

$$N_0^2 = N_a(C + N_a) \tag{6.45}$$

and

$$N_a = \frac{C}{2}\left[\left(1 + \frac{4N_0^2}{C}\right)^{1/2} - 1\right] \tag{6.46}$$

$$N_c = \frac{C}{2}\left[\left(1 + \frac{4N_0^2}{C}\right)^{1/2} + 1\right] \tag{6.47}$$

If $C \ll N_0$, diffusion is controlled by the thermally activated formation and migration of vacancies and the activation enthalpies are given by (6.42) and (6.43) (*intrinsic regime*).

If $C \gg N_0$, the concentration of thermal vacancies is negligible compared to that of the charge compensating vacancies fixed by the concentration of impurities:

$$N_c \cong C \tag{6.48}$$

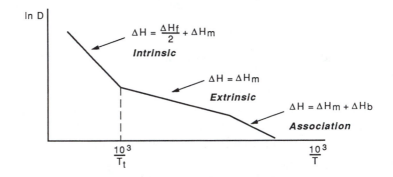

Figure 6.3. Diffusion regimes for ionic crystals: At high temperature, the concentration of vacancies is the thermal equilibrium concentration; below a temperature T_t that depends on the concentration of aliovalent impurities, the concentration of vacancies is imposed by charge balance requirements; at still lower temperatures, the charge-compensating vacancies remain associated to the impurities and have to be dissociated to participate in diffusion. ΔH_f is the formation energy, ΔH_m is the migration energy, and ΔH_b is the binding energy of the vacancies.

The cationic vacancies are freely available for diffusion, and no formation activation enthalpy has to be spent:

$$\Delta H_c = \Delta H_{mc} \tag{6.49}$$

Conversely, the concentration of anionic vacancies is reduced, $N_a = N_0^2/C$, thus reducing the diffusion coefficient of the anions (*extrinsic regime*). At lower temperatures, the charge-compensating vacancies may remain associated with the aliovalent cations; the activation enthalpy for diffusion then includes the dissociation energy (equal to the binding energy) of the complex (*association regime*).

On an Arrhenius plot (Fig. 6.3), the intrinsic and extrinsic regimes are represented by straight line segments with different slopes and the transition temperature increases with impurity content:

$$T_t = \Delta H_f (\Delta H_f - 2k \ln C)^{-1} \tag{6.50}$$

In the case of compounds where a cation has several oxidation states (as often happens with transition metals oxides and silicates), the concentration of vacancies is sensitive to the oxygen partial pressure. For instance, in a binary oxide such as magnesiowüstite $(Mg, Fe)O$, Fe^{2+} ions can be oxidized to Fe^{3+} and we have (Chen and Peterson 1980)

$$\tfrac{1}{2}O_g + 2Fe^{2+} \rightarrow O^{2-} + V_M^{2'} + 2Fe^{3+} \tag{6.51}$$

$V_M^{2'}$ represents an Fe^{2+} vacancy, with effective charge $2-$, an electron hole h is associated with an Fe^{3+} ion. The law of mass action gives

$$[V_M^{2'}][h]^{1/2}(p_{O_2}^2)^{-1/2} = K \tag{6.52}$$

and electrical neutrality imposes the condition $2[V_M^{2'}] = [h]$; hence, for a diffusion controlled mechanism,

$$D_M \propto [V_M^{2'}] \propto (p_{O_2}^2)^{1/6} \tag{6.53}$$

Nonstoichiometry, which in turn is often a function of oxygen partial pressure, obviously considerably complicates the problem (see Kofstad 1983; Dieckmann 1984).

The transport of matter by diffusion in a compound involves transport of all cations and anions corresponding to the formula group. The case of binary ionic crystals $A_\alpha B_\beta$ has been treated by Ruoff (1965) (for a summary see Poirier 1985), who defined an effective diffusion coefficient

$$D_{eff} = \frac{D_A D_B}{n_A D_B + n_B D_A} \tag{6.54}$$

where D_A, n_A and D_B, n_B are the diffusion coefficients and atomic fractions of species A and B, respectively ($n_A = \alpha/(\alpha+\beta)$, $n_B = \beta/(\alpha+\beta)$). The case of multicomponent diffusion in silicates has been treated by Lasaga (1979).

Values of the preexponential coefficient D_0 and of the activation enthalpies for self-diffusion or hetero-diffusion, from recent experiments on various minerals, are given in Table 6.1. See also Freer (1980) for a review on oxides and Freer (1981) for a review on silicates. Experimental methods of determination of the diffusion coefficients are reviewed in Ryerson (1987). For more details on the mechanisms of diffusion, the reader is referred to the book by Borg and Dienes (1988).

i. Empirical relations

i. A systematic empirical proportionality relation, the *van Liempt relation,* has been found between the activation enthalpy of self-diffusion and the melting point of metals (van Liempt 1935; Bocquet, Brébec, and Limoge 1983):

$$\Delta H_d = 34 T_m \tag{6.55}$$

with ΔH_d in calories per mole and T_m in Kelvins (Fig. 6.4). In other words, the activation enthalpy of self-diffusion of metals scales with their melting points, and the self-diffusion coefficient can be written

Table 6.1. *Preexponential factor and activation energy for the diffusion in a few minerals*

System	ΔT (°C)	D_0 (cm^2/s)	ΔH (kJ/mol)	Reference
^{30}Si in Forsterite	1300–1700	1.5×10^{-6}	376 ± 42	Jaoul et al. (1981)
^{18}O in Forsterite	1275–1625	3.5×10^{-3}	372 ± 13	Reddy et al. (1980)
^{18}O in Forsterite	1300–1600	2.3×10^{-6}	293 ± 42	Jaoul, Houlier, and Abel (1983)
^{18}O in Forsterite	1472–1734	2.9×10^{-2}	416	Oishi and Ando (1984)
^{18}O in Magnetite ($P_{\text{Water}} = 1$ kbar)	550–800	3.5×10^{-6}	188	Giletti and Hess (1988)
^{85}Sr in Diopside (diffusion $\parallel c$)	1100–1300	54	405	Sneeringer, Hart, and Shimizu (1984)
^{18}O in β-Quartz ($P_{\text{Water}} = 1$ kbar) (diffusion $\parallel c$)	600–800	4×10^{-7}	142 ± 4	Giletti and Yund (1984)
(diffusion $\perp c$)		1×10^{-4}	234 ± 8	
(diffusion $\parallel c$)	700–850	2.1×10^{-7}	139	Dennis (1984)
(diffusion $\perp c$)		3.2×10^{-6}	204	

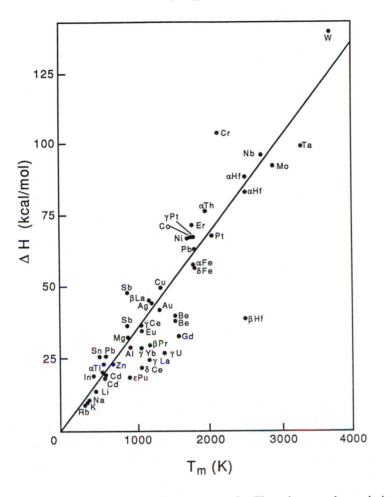

Figure 6.4. Van Liempt relation for some metals: There is a good correlation between the activation energy for diffusion ΔH and the melting temperature T_m (after Bocquet et al. 1983).

$$D = D_0 \exp\left(\frac{-g_d T_m}{T}\right) \qquad (6.56)$$

with $g_d = 17$. Hence

$$\Delta H_d = g_d R T_m \qquad (6.57)$$

The van Liempt relation does not hold for oxides and silicates: Plots of published values of activation enthalpies for the self-diffusion of oxygen or cations (Freer 1980, 1981) against the melting temperatures of oxides

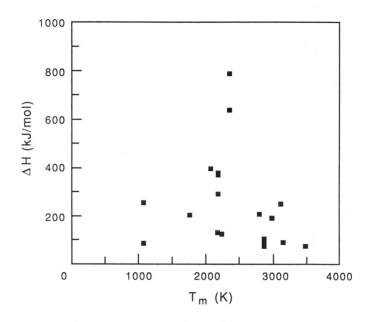

Figure 6.5. Activation energy for the diffusion at high temperature of ^{18}O in binary oxides versus melting point (data from Freer 1980). The van Liempt relation is not verified.

or silicate minerals yield only a scattered cloud of points (Fig. 6.5 shows for instance all the measured activation enthalpies of ^{18}O in binary oxides). This is obviously due to the fact that diffusion in these compounds is often extremely sensitive to the impurity content, the oxygen partial pressure, and the experimental temperature range; in most cases the published values are not comparable.

ii. The *compensation law* or "isokinetic effect" was recently revived by Hart (1981) and applied to diffusion in minerals. It is an empirical linear correlation between the logarithm of the preexponential factor D_0 and the activation enthalpy for the diffusion of various elements in the same mineral or rock (Fig. 6.6). For olivine, for instance, Hart (1981) found (Fig. 6.7)

$$\Delta H = 78 + 7.5 \log D_0 \qquad (6.58)$$

This correlation implies that there exists an "isokinetic" temperature T^* for which all coeffiicients of diffusion are equal to D^*. Then

$$\Delta H = RT \ln D_0 - RT^* \ln D^* \qquad (6.59)$$

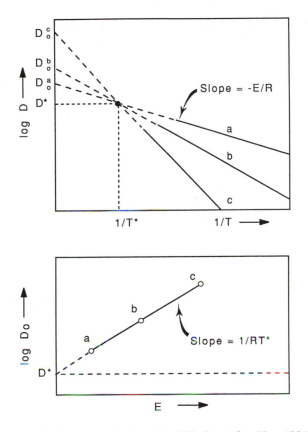

Figure 6.6. Compensation law for diffusion (after Hart 1981).

and, as D_0 is proportional to $\exp(\Delta S/R)$, this in fact is equivalent to having a linear relation between the activation enthalpy and entropy. The compensation law, also applied to other thermally activated phenomena, such as the thermal death of bacteria, has a long history of controversy (Exner 1964; Banks, Damjanovic, and Vernon 1972; Boon 1973; Harris 1973; Kemeny and Rosenberg 1973). The case for the correlation being in most cases devoid of any physical significance has convincingly been made (Exner 1964; Banks et al. 1972; Dosdale and Brooks 1983; Kirchheim and Huang 1987): It could be due to logarithmic compression in plotting log D, which varies in a rather small range, since diffusion to be measurable has to be neither too rapid nor too slow. In other words, only those materials that apparently obey the compensation law are susceptible to investigation (Banks et al. 1972).

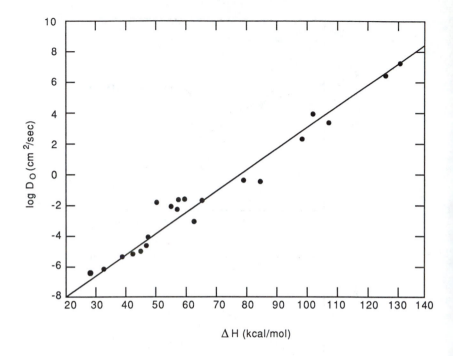

Figure 6.7. Compensation law for diffusion in a few olivines (Mg–Fe, Co, Ni, Mn) (after Hart 1981).

ii. Effect of pressure

The pressure dependence of the diffusivity is expressed by the *activation volume,* defined as

$$\Delta V = \frac{\partial \Delta G}{\partial P} \tag{6.60}$$

ΔG is the Gibbs free energy of the activation process; the apparent activation volume, defined by (6.31), $\Delta V' = -RT(\partial \ln D/\partial P) = \partial \Delta H/\partial P$, is equal to the real activation volume only if the pressure dependence of the entropic term is neglected.

As diffusion experiments under pressure are difficult to carry out, it is interesting to calculate or estimate the activation volume, on theoretical grounds or from empirical relations.

Keyes (1958, 1960, 1963) proposed a semi-empirical relation

$$\Delta V' \cong 4K^{-1}\Delta H \tag{6.61}$$

where $\Delta V'$ is the apparent activation volume and K is the isothermal bulk modulus. He found it could be justified using several models of lattice defects. Assuming that the free energy of formation and migration of the defect responsible for the diffusion is essentially represented by strain energy in the crystal considered as a continuous medium, we can write

$$\Delta G \propto \mu V \qquad (6.62)$$

or

$$\Delta G \propto KV \qquad (6.63)$$

depending on whether we assume the energy is entirely due to shear or dilatation. From the definition (6.60), we can write

$$\frac{\Delta V}{\Delta G} = \frac{\partial \ln \Delta G}{\partial P}$$

If we assume that the pressure dependence of the bulk and shear moduli are equal, we can write, using Slater's relation (3.65 and 3.66),

$$\frac{\partial \ln \mu}{\partial P} = \frac{\partial \ln K}{\partial P} = \frac{2\gamma + \frac{1}{3}}{K} \qquad (6.64)$$

and, whether the energy is due to shear or dilatation,

$$\Delta V = [(2\gamma + \tfrac{1}{3})K^{-1} - K^{-1}]\Delta G$$

$$\Delta V = 2(\gamma - \tfrac{1}{3})K^{-1}\Delta G \qquad (6.65)$$

which is equivalent to equation (6.61) if we take $\gamma = 1.83$, corresponding to $K' = 4$ (see Sec. 4.4), and if we assume that $\Delta G \cong \Delta H$.

Sammis, Smith, and Schubert (1981) have shown that strain energy models assuming pure shear and pure dilatation give upper and lower bounds on the activation volume.

Keyes (1963) also considered the case where the activation enthalpy is expressed in terms of the melting temperature (equation 6.57), the apparent activation volume is then, using Clapeyron's relation,

$$\Delta V' = \frac{dT_m}{T_m \, dP}\Delta H = \frac{\Delta V_m}{T_m \Delta S_m}\Delta H = \frac{\Delta V_m}{\Delta H_m}\Delta H \qquad (6.66)$$

where T_m, ΔV_m, and ΔS_m are the melting temperature, volume, and entropy. The relation can also be written

$$\Delta V' = \frac{d \ln T_m}{dP}\Delta H = \frac{d \ln T_m}{d \ln \rho}K^{-1}\Delta H$$

and using the differential Lindemann law (5.26), we obtain (Poirier and Liebermann 1984)

$$\Delta V' = 2(\gamma - \tfrac{1}{3})K^{-1}\Delta H \qquad (6.67)$$

which is equivalent to (6.65) if $\Delta V' \cong \Delta V$ and to (6.61) if $\gamma = 1.83$.

At high pressures, the assumption that the activation volume is constant does not hold any longer. The pressure dependence of the activation volume has been calculated, using various assumptions, by O'Connell (1977), Karato (1981), and Sammis et al. (1981). It can be expressed as a dimensionless parameter (Poirier and Liebermann 1984).

Differentiating (6.65) with (6.60), we obtain

$$\frac{\partial \ln \Delta V}{\partial P} = \frac{\partial \ln[2K^{-1}(\gamma - \tfrac{1}{3})]}{\partial P} + \frac{\Delta V}{\Delta G} \qquad (6.68)$$

and

$$\frac{\partial \ln \Delta V}{\partial P} = \frac{\partial \ln[2K^{-1}(\gamma - \tfrac{1}{3})]}{\partial P} + \frac{2(\gamma - \tfrac{1}{3})}{K}$$

which, with (6.64), gives

$$\frac{\partial \ln \Delta V}{\partial P} = \frac{1}{K} + \frac{1}{\gamma - \tfrac{1}{3}}\frac{\partial \gamma}{\partial P} \qquad (6.69)$$

Recalling that $\gamma \rho = $ const. (3.85), one obtains

$$\frac{\partial \ln \Delta V}{\partial \ln \rho} = -1 + \gamma\left(\gamma - \frac{1}{3}\right)^{-1} \qquad (6.70)$$

Experimental and calculated values of activation volumes for self-diffusion can be found in Sammis et al. (1981).

6.3 Viscosity of solids

The viscosity of solids is defined, as for fluids, by equation (6.7), rewritten here:

$$\eta = \frac{\sigma}{\dot{\epsilon}} \qquad (6.71)$$

where σ is the applied shear stress and $\dot{\epsilon} = d\epsilon/dt$ is the shear strain rate (Fig. 6.1). The definition of viscosity as a material constant has a meaning only if it does not depend on time, that is, if the shear strain rate is constant. This is the case for *high-temperature creep* under constant stress, in the quasi-steady-state regime. The problem of the viscosity of solids is therefore that of the high-temperature creep of crystals. A summary overview of the principal physical mechanisms of creep is given in what follows and, for further information, the reader is referred to the monograph by Poirier (1985). If the viscosity depends only on the structural properties of the material and on temperature, it is said to be *Newtonian;* the creep rate $\dot{\epsilon}$ then depends linearly on stress. In many cases, however, the creep rate is an increasing nonlinear function of stress, usually fitted by a power law, $\dot{\epsilon} \propto \sigma^n$, with $3 \le n \le 5$ for most solids (*"power-law creep"*).

Creep of solids is a thermally activated phenomenon, with an activation energy ΔH, and the creep rate can be put in the Arrhenian form

$$\dot{\epsilon} = \dot{\epsilon}_0 \sigma^n \exp\left(\frac{-\Delta H}{RT}\right) \tag{6.72}$$

To define the viscosity in the non-Newtonian case, one must therefore specify either the stress or the strain rate (see, e.g., Poirier 1988b).

The viscosity at constant stress is given by

$$\eta_\sigma \equiv \frac{\sigma}{\dot{\epsilon}(\sigma)} = \dot{\epsilon}_0^{-1} \sigma^{1-n} \exp\left(\frac{\Delta H}{RT}\right) \tag{6.73}$$

and the viscosity at constant strain rate by

$$\eta_{\dot{\epsilon}} \equiv \frac{\sigma(\dot{\epsilon})}{\dot{\epsilon}} = \dot{\epsilon}_0^{-1/n} \sigma^{(1-n)/n} \exp\left(\frac{\Delta H}{nRT}\right) \tag{6.74}$$

It is also interesting to consider the viscosity at constant dissipated power ($\sigma\dot{\epsilon} = \text{const.}$):

$$\eta_{\sigma\dot{\epsilon}} \equiv (\eta_\sigma \eta_{\dot{\epsilon}}^n)^{1/(n+1)} = \dot{\epsilon}_0^{-2(n+1)} \sigma^{(1-n)/(1+n)} \exp\left[\frac{2\Delta H}{(n+1)RT}\right] \tag{6.75}$$

Note that the apparent activation energies (i.e., the sensitivity to temperature) of the viscosity at constant strain rate or dissipated power are smaller than that at constant stress, unless, of course, if $n = 1$, in which case all three viscosities are equal.

Creep deformation is due to a transport of matter by motion of lattice defects. If the defects are vacancies, creep, then, directly results from the directed diffusion of matter and vacancies in opposite senses, and it is Newtonian. If the defects are dislocations, creep is non-Newtonian and can be controlled either by diffusion-controlled climb or by glide of dislocations (see subsequent subsections).

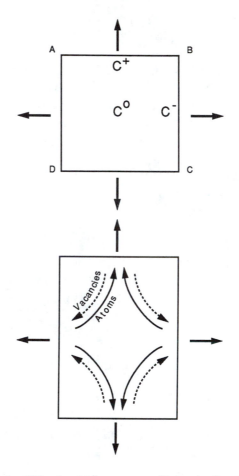

Figure 6.8. Principle of Herring–Nabarro creep. Vacancies flow from the faces in tension (concentration C^+) to the faces in compression (concentration $C^- < C^+$) and matter flows in the opposite direction.

i. Diffusion creep

Diffusion creep, or *Nabarro–Herring creep,* is due to transport of matter by self-diffusion through the grains of a polycrystal.

Let us consider a crystal of size d in the shape of a cube (for the sake of simplicity), subjected to a normal compressive stress on a pair of faces and to normal tensile stresses on the four other faces (Fig. 6.8). The compressive normal stress hinders the formation of vacancies (to form vacancies, atoms are extracted from inside the crystal and deposited on the surface), while the tensile stress facilitates it. The formation energy of vacancies is therefore increased by $\sigma V/kT$ on the faces in compression and decreased by the same amount on the faces in tension (V is the atomic volume) and the equilibrium concentrations of vacancies on the faces in compression and tension are, respectively,

$$C^- = C_0 \exp\left(\frac{-\sigma V}{RT}\right) \tag{6.76}$$

$$C^+ = C_0 \exp\left(\frac{\sigma V}{RT}\right) \tag{6.77}$$

There is a flow of vacancies from the faces in tension to the faces in compression and a flow of atoms in the opposite direction. The flux of atoms is given by Fick's law:

$$\mathbf{J} = -D_v \operatorname{grad} C = -\frac{\alpha D_v(C^+ - C^-)}{d} \tag{6.78}$$

where D_v is the diffusion coefficient of vacancies and α is a geometrical constant. In the unit time, a number Jd^2 of atoms leaves the faces in compression and are added to the faces in tension; the crystal shortens by Δd and widens by the same quantity. We have

$$\Delta d = \frac{(Jd^2)V}{d^2} = JV$$

Hence

$$\dot{\epsilon} = \frac{\Delta d}{d} = \frac{JV}{d} \tag{6.79}$$

With (6.76), (6.77), and (6.78), we obtain

$$\dot{\epsilon} = \alpha D_v C_0 V d^{-2} \sinh\left(\frac{\sigma V}{k_B T}\right) \tag{6.80}$$

or, remembering that $D_v C_0 V = D_v N_v = D_{sd}$, the coefficient of self-diffusion, we obtain, for $\sigma V \ll k_B T$,

$$\dot{\epsilon} = \frac{\alpha D_{sd} \sigma V}{d^2 k_B T} \tag{6.81}$$

The activation energy of the viscosity is therefore obviously equal to the activation energy of the self-diffusion.

ii. Dislocation creep

Dislocations (Fig. 6.9) are line defects bounding an area within the crystal where slip by an interatomic distance b has taken place; that is, the crystal on one side of the slip plane has been rigidly displaced by b with respect to the other side (see Friedel 1964, and, for a short summary, Poirier 1985). Dislocations move under an applied stress, extending the slipped area. When a dislocation loop sweeps a whole crystal plane and moves out of the crystal, it leaves at the surface a step of height b. The crystal has acquired a permanent shear deformation equal to b/h, where h is the dimension of the crystal normal to the slip plane (Fig. 6.10). The motion of dislocations by progressive breaking and reestablishment of atomic bonds is the energetically economic way of deforming crystals. Dislocations are carriers of plastic deformation in solids.

For a mobile dislocation density ρ (total length of dislocation lines per unit volume) moving an average distance ΔL, the shear strain is

$$\epsilon = \rho b \Delta L \tag{6.82}$$

and, for constant dislocation density, the creep rate is

$$\dot{\epsilon} = \rho b \bar{v} \tag{6.83}$$

where $\bar{v}$ is the average dislocation velocity. This is *Orowan's equation,* which can be seen as a transport equation, giving the creep rate as the product of the density of strain carriers (the dislocations), their strength (the *Burgers vector* **b**, characteristic of the dislocation and equal to the elementary slip by an interatomic distance), and their velocity. The equation is analogous to the microscopic Ohm's law (see Sec. 6.5). The dependence of the viscosity on stress and temperature is found by expressing the dependence of dislocation density and velocity on them.

For steady-state creep, the dislocation density is generally proportional to σ^2 (see Poirier 1985) and depends little on temperature. The average dislocation velocity depends of course on σ and is controlled by the nature and distribution of obstacles to the motion of dislocation lines in the crystal. Two cases are especially interesting:

a. The principal obstacle to the dislocation motion lies in the intrinsic difficulty in breaking the atomic bonds, the so-called lattice friction. The material is said to have a high *Peierl's stress* and the dislocation lines tend to be straight, lying in potential valleys. The lattice friction can be overcome with the help of thermal vibrations and of the effective stress (applied stress minus internal stress). The motion is directly

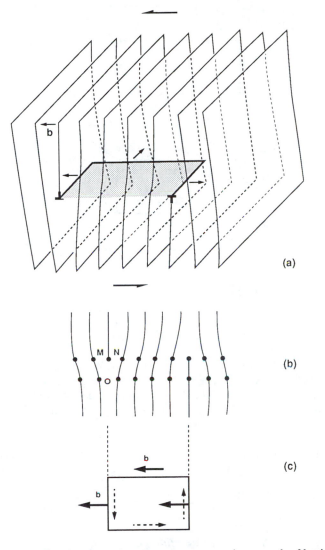

Figure 6.9. (a) Dislocation loop in a crystal represented as a stack of lattice planes (the front half is removed for clarity). The slipped area inside the dislocation is stippled. The extra half-plane above the edge portion (left) is seen to wind itself in helicoidal fashion around the screw portion (back) until it ends up as the extra half-plane below the other edge portion (right). Each turn of the screw corresponds to a displacement by the Burgers vector **b**, parallel to the screw portion and normal to the edge portions. (b) Front view of the half loop. Flipping the bond *OM* to *ON* makes the edge portion propagate toward the left. (c) Top view of the whole loop oriented continuously (dashed arrows); opposite-sign extra half-planes correspond to Burgers vectors of edge portions that would point in the opposite direction if the dislocation segments were oriented in the same direction.

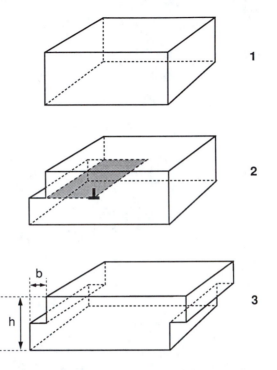

Figure 6.10. Slip by dislocation motion. (1) Undeformed crystal. (2) A dislocation line has been created at left and moved inside the crystal, causing a slip by b
behind it (stippled area). (3) The dislocation has swept the whole length of the
crystal and left a step of height b at the surface. The shear strain is $\epsilon = b/h$.

thermally activated and the activation enthalpy is stress dependent,
decreasing with increasing stress. The creep rate can be written

$$\dot{\epsilon} = \dot{\epsilon}_0 \sigma^2 \exp\left[\frac{-\Delta H(\sigma)}{RT}\right] \tag{6.84}$$

The viscosity is said to be *glide-controlled.*
b. If the obstacles are discrete, with an average spacing ΔL, the average
velocity can be written, in general, as

$$\bar{v} = \frac{\Delta L}{t_g + t_o} \tag{6.85}$$

where t_g is the average time the dislocation takes in gliding over the
distance ΔL between obstacles and t_o is the average time it takes in
overcoming the obstacles. If $t_g \ll t_o$, the average velocity is $\bar{v} = \Delta L/t_o$.
Edge dislocations can overcome obstacles to glide by moving out of

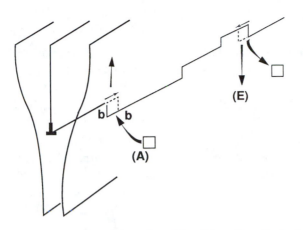

Figure 6.11. Climb of an edge dislocation. The dislocation climbs by an inter-atomic distance when a jog travels down its length by absorbing (A) or emitting (E) vacancies.

their glide plane by diffusion-controlled climb (Fig. 6.11) (*Weertman creep*). If δ is the distance a dislocation has to climb to escape the obstacle and v_c is the climb velocity, we have

$$\dot\epsilon = \rho b \Delta L v_c \delta^{-1} \tag{6.86}$$

It can be shown (Friedel 1964) that v_c is proportional to the self-diffusion coefficient D_{sd} and, in the linear approximation, to the applied stress σ. Remembering that $\rho \propto \sigma^2$, we then obtain the general equation for diffusion-controlled creep:

$$\dot\epsilon = \dot\epsilon_0 \sigma^3 \exp\left(\frac{-\Delta H_{sd}}{RT}\right) \tag{6.87}$$

We have here a physical justification for power-law creep with $n = 3$, but the experimental values usually range between 3 and 5. Diffusion-controlled power-law creep is often observed at high temperatures in metals and minerals, although some caution is in order before issuing general pronouncements (Poirier 1978). It is currently thought to be responsible for the viscosity of the Earth's mantle (Weertman 1970, 1978; Poirier 1988b). The viscosity profile of the mantle then depends on the temperature profile and on the pressure dependence of the creep rate.

iii. Effect of pressure

As for diffusion, the pressure dependence is expressed by an apparent activation volume

$$\Delta V' = RT \frac{\partial \ln \eta}{\partial P} = \frac{\partial \Delta H_v}{\partial P} \qquad (6.88)$$

which should in principle be equal to the activation volume for self-diffusion of the atomic species whose diffusion controls the creep rate. There is, however, some debate as to which species controls the creep rate, even in such well-investigated minerals such as olivine (e.g., Jaoul et al. 1981) and, anyway, there are no experimental determinations of the activation volumes. One must therefore resort to systematics.

Weertman (1970) applied van Liempt's relation (6.57) to the creep of metals and ice and implicitly extended it to the creep of all substances, including the minerals of the Earth's lower mantle. Weertman's relation,

$$\Delta H_v = g_v RT_m \qquad (6.89)$$

where ΔH_v is the activation energy for viscosity (or creep rate) and T_m is the melting temperature, is valid for metals with a value of the constant g_v close to 18. For compounds, however, things are obviously not so simple since the creep mechanisms are not always clear and since even if creep is diffusion-controlled, van Liempt's relation does not hold. Besides, the published values of activation enthalpies for creep of ceramics (Cannon and Langdon 1983) correspond to widely differing experimental conditions, purity of materials, grain size, oxygen partial pressure, and so forth. Even if the data are carefully selected, a plot of the activation energy against melting temperature for various materials exhibits a considerable scatter (Fig. 6.12 shows, for instance, the data corresponding to compression creep of single crystals of oxides and alkali halides). By grouping the compounds according to their structure, it is, however, possible to improve the systematics (Frost and Ashby 1982) (Fig. 6.13). Nevertheless, there is little hope of ever obtaining a systematics good enough to usefully predict the activation enthalpy of a compound from the knowledge of its melting temperature because the scatter on g_v remains much too large. We find, for instance, that, from the data of Figure 6.12, we have $g_v = 27 \pm 8$ for oxides (in agreement with Frost and Ashby 1982), but since g_v appears in an exponential, this corresponds to an uncertainty of three orders of magnitude on the viscosity! Weertman's relation, however, remains useful for the purpose originally assigned to it by Weertman (1970), that is, to extrapolate the viscosity to high pressures, if the melting curve is known or inferred, but the value of g_v must be the one experimentally determined for the specific compound one is interested in.

The activation volume for diffusion-controlled creep of solids (and the Earth's mantle) was estimated by several authors (e.g., Karato 1981b;

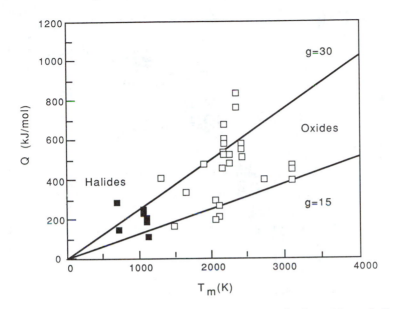

Figure 6.12. Activation energy for creep versus melting point for oxides and alkali halides (data from Cannon and Langdon 1983). Straight lines are drawn for $g = 15$ and $g = 30$.

Sammis et al. 1981; Ellsworth, Schubert, and Sammis 1985), using either Weertman's relation (6.88), that is, assuming that the pressure dependence of viscosity is equal to that of the melting point, or Keyes's formulation with an elastic strain model for the Gibbs free energy of activation. Poirier and Liebermann (1984) discussed these formulations and showed that they were equivalent within the limits of validity of Slater's relation and Lindemann's law. The discussion has already been presented for diffusion in Section 6.2.

6.4 Diffusion and viscosity in liquid metals

The Earth's outer core (16 percent of the Earth's volume) is constituted of liquid iron. Its diffusional and viscous transport properties must be replaced in the wider framework of liquid metals, which, being simple unassociated liquids, have been experimentally and theoretically well investigated. The reader is referred to the recent book by Shimoji and Itami (1986) for a complete review of the field (see also Battezzati and Greer 1989).

There is an intimate connection in liquids between the self-diffusion coefficient and the viscosity η. It is expressed by the *Stokes–Einstein relation*

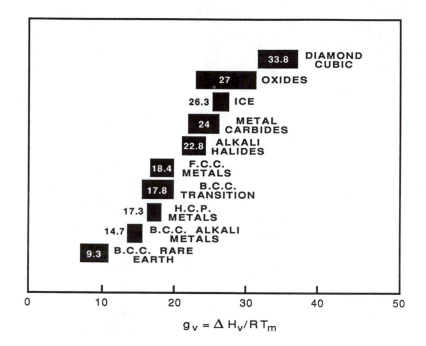

Figure 6.13. The ranges of the dimensionless quantity $g_v = \Delta H/RT_m$ for classes of solids (after Frost and Ashby 1982).

$$D\eta = \frac{k_B T}{a} \qquad (6.90)$$

where a is a parameter linked to the atomic size. The Stokes–Einstein relation has been experimentally verified for liquid metals (Saxton and Sherby 1962) and can be theoretically justified (e.g, Zwanzig 1983).

Eyring (1936) considers a liquid as a collection of molecules bound to their neighbors and of "dissolved holes" (similar to vacancies in solids). Transport, either by diffusion or under the action of a shear stress (viscosity), is thought to occur by thermally activated jumps of the molecules into the holes. The diffusion coefficient then can be written, as for solids (see 6.36),

$$D = \lambda^2 k' \qquad (6.91)$$

where λ is the jump distance, comparable to the average interatomic spacing, and k' is the absolute reaction rate of the diffusion process. Eyring therefore finds an expression of the diffusion coefficient that follows an Arrhenius law, like the coefficient of diffusion of crystals:

$$D = D_0 \exp\left(\frac{-Q}{RT}\right) \tag{6.92}$$

where Q is the activation energy corresponding to the potential barrier that the molecule must overcome to jump into the hole.

To calculate the viscosity, Eyring assumed that the motion takes place by individual molecules slipping by a distance λ over the potential barrier with a reaction rate k'. The viscosity is given by equation (6.17) with $\dot{\epsilon} = \lambda_1/\Delta v$, where Δv is the difference in velocity of two neighboring parallel layers, which are a distance λ_1 apart. The force acting on a molecule is $\sigma \lambda_2 \lambda_3$, where λ_2 and λ_3 are the distances between molecules in the parallel planes in the direction of motion and perpendicular to it, respectively. The force acts to lessen the work for overcoming the barrier in the forward sense and raises it in the backward sense, so the velocity difference can be written

$$\Delta v = \lambda k' \left[\exp\left(\frac{\sigma \lambda_2 \lambda_3 \lambda}{2 k_B T}\right) - \exp\left(\frac{-\sigma \lambda_2 \lambda_3 \lambda}{2 k_B T}\right) \right] \tag{6.93}$$

Now, for $\sigma \lambda_2 \lambda_3 \lambda \ll 2 k_B T$, we have

$$\eta = \lambda_1 k_B T (\lambda^2 \lambda_2 \lambda_3 k')^{-1} \tag{6.94}$$

With (6.91) and (6.94), we see that Eyring's theory yields the Stokes–Einstein relation

$$D\eta = \frac{\lambda_1}{\lambda_2 \lambda_3} k_B T \tag{6.95}$$

The "free volume" theory of diffusion of Cohen and Turnbull (1951) assumes that atoms are enclosed in cages formed by their neighbors and can jump only when statistical fluctuations of the free volume V_f (total volume minus the volume occupied by the atoms) create a hole of volume V^* sufficient to accept an atom. The diffusion coefficient is

$$D \propto (k_B T)^{1/2} \exp\left(\frac{-\gamma V^*}{V_f}\right) \tag{6.96}$$

where $0 < \gamma < 1$ is a numerical constant and $V_f \propto \alpha(T - T_0)$, where α is the coefficient of thermal expansion.

Swalin (1959) starts from Einstein's equation (6.36) but, instead of considering atoms jumping by a fixed distance into holes of a critical size, he assumes that local density fluctuations cause atoms to move small variable distances and finds for the diffusion coefficient of liquid metals

$$D = (1.29 \times 10^{-8}) \frac{\beta^{-2}T^2}{\Delta H_v} \tag{6.97}$$

where β (in Å^{-1}) is related to the curvature of the potential versus distance curve and ΔH_v is the heat of vaporization. In this model, there is a distribution of fluctuation sizes and most atoms participate in the diffusion process at a given moment. The process has no activation energy; however, a plot of $\ln D$ versus $1/T$ yields an apparent "activation energy" equal to $2RT$, devoid of physical meaning.

Indeed, the assumption that there are identifiable holes and that there is a localized energy barrier to diffusion, inherent to Eyring's "activated state" model and its variants, is difficultly tenable in liquids.

Nachtrieb (1967, 1977) found that the experimental results were best fitted by a linear relation between D and T, based on the assumption that each atom vibrates in the cage of its neighbors with a harmonic motion and that with each oscillation, the center of oscillation moves by a distance equal to the mean thermal amplitude

$$D = \frac{k_B \Theta_D}{h\kappa} T \tag{6.98}$$

where Θ_D is the Debye temperature, h is Planck's constant, and κ is the force constant of the harmonic vibrations. In a sense, the entire liquid is in an activated state and the concept of activation energy has no meaning. However, despite the fact that there is no convincing theoretical justification for it, the temperature dependence of the diffusion coefficient is still currently displayed on Arrhenius plots and "activation energies" are determined, even by those who object to it (Nachtrieb 1967; Shimoji and Itami 1986). Indeed, the fit of the data by an Arrhenius law is in most cases reasonably good and "activation energies" provide a convenient means of systematizing the data.

The "activation energies" for self-diffusion and heterodiffusion in liquid metals are, not surprisingly, much smaller than in the corresponding solids. As an example, average values for the heterodiffusion of H, C, N, O, and S in α, γ, and liquid iron, from von Horst Bester and Lange (1972), are given in Table 6.2.

As for diffusion, Eyring's (1936) activated state theory of viscosity is among the very few that predict an Arrhenian temperature dependence (see Brush (1962) and Shimoji and Itami (1986) for a review).

Andrade (1934, 1952) proposed a theory of the viscosity of unassociated liquids at their melting points, based on the idea that their structure is close to that of the solid and that their viscosity is due to the transfer of

Table 6.2. *Apparent activation energies for diffusion of various elements in* α, γ, *and liquid iron*

Diffusing element	$\bar{Q}_\alpha$ (kJ/mol)	$\bar{Q}_\gamma$ (kJ/mol)	$\bar{Q}_{liq}$ (kJ/mol)
Hydrogen	12.5	47.2	15.5
Carbon	81.3	134.4	49
Nitrogen	76.8	152.2	50.2
Oxygen	96.3	168.5	50.2 ± 8.4
Sulfur	207.2	222.4	35.6

Source: von Horst Bester and Lange (1972).

momentum by collisions between parallel layers of atoms through the amplitude of thermal vibration. Andrade found an expression of the viscosity in terms of the mass m of atoms, the mean interatomic spacing, and the Lindemann vibrational frequency ω_L (equation (5.19)):

$$\eta = 2m\omega_L(3\pi a)^{-1} \qquad (6.99)$$

or, in cgs units,

$$\eta = 5.7 \times 10^{-4}(AT_m)^{1/2}V^{-2/3} \qquad (6.100)$$

where A and V are the atomic mass and volume. Viscosity is thought to decrease with increasing temperature because the thermal agitation interferes with the transfer of momentum at maximum amplitude. The variation of viscosity with temperature is then governed, according to Andrade (1934), by the fraction of atoms possessing the mutual potential energy at extreme amplitude, which varies according to the Boltzmann distribution formula, hence

$$\eta \propto \exp\left(\frac{C}{T}\right) \qquad (6.101)$$

where C has the dimensions of energy. Although it is not an activation energy, it can be treated as an apparent "activation energy" for systematization purposes.

Empirical relations and effect of pressure

As for diffusion and viscosity in solids, there are systematic empirical relations between the apparent "activation energies" of diffusion and viscosity in liquid metals similar to (6.57) and (6.89), but the values of the

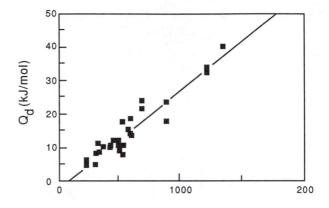

Figure 6.14. Correlation between the melting point and apparent activation energy
for self-diffusion of liquid metals (after Poirier 1988).

constants g_d and g_v are much smaller (Saxton and Sherby 1962; Poirier
1988a). This can be expected since the activation energy for diffusion (and
diffusion-controlled creep) of solids consists mostly of the formation energy of point defects, whereas in liquids the apparent "activation energy"
only reflects the temperature dependence of the mobility of atoms, which
does not necessitate the creation of localized defects.

Poirier (1988a) found for a number of liquid metals

$$Q_d = 3.2RT_m \tag{6.102}$$

for diffusion (Fig. 6.14) and

$$Q_v = 2.6RT_m \tag{6.103}$$

for viscosity (Fig. 6.15). The value of $g_d \cong 3$ is in good agreement with
Saxton and Sherby (1962) and with Nachtrieb (1967), whose model for
diffusion rests on the idea that the average energy of atoms vibrating in
the cage of their neighbors is equal to $3RT$.

The extrapolation to high pressures of the viscosity of liquid metals
(especially interesting for iron in the Earth's core) stands on somewhat
safer ground than that for solids, despite the lack of theoretical justification for equation (6.103), because the value of the constant (close to 3) has
less scatter. There is an excellent agreement between the apparent "activation volume" of the viscosity of mercury, calculated from the experimen-
tal results of Bridgman, using equation (6.88) ($\Delta V' = 0.62 \text{ cm}^3/\text{mol}$), and
the empirical value, calculated from equation (6.103) and the slope of the
melting curve ($\Delta V' = 0.59 \text{ cm}^3/\text{mol}$).

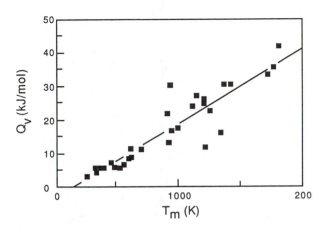

Figure 6.15. Correlation between the melting point and apparent activation energy for viscosity of liquid metals (after Poirier 1988).

It is interesting to consider the viscosity of a liquid metal right at its melting point $(T = T_m)$. Since we have

$$\eta \propto \exp\left(\frac{g_v T_m}{T}\right) \qquad (6.104)$$

we immediately see that the viscosity stays constant along the melting curve and equal to the viscosity at atmospheric pressure. The viscosity of liquid iron at the inner core boundary should then be equal to 6 centipoises (Poirier 1988a).

6.5 Electrical conduction

6.5.1 Generalities on the electronic structure of solids

The theory of the electronic structure of solids is the cornerstone of solid-state physics, and it is obviously beyond the scope of this book even to give a summary overview of it. However, since transport of electronic charges is what electrical conduction is all about, it is impossible to avoid giving some indispensable background, mostly at the hand-waving level. For an elementary or intermediate treatment, the reader is referred to Ziman (1965), Kittel (1967), Honig (1970), or Animalu (1977).

An essential concept in solid-state theory is that of electronic *energy bands,* separated by *band gaps.* Energy bands are ranges of allowed energy

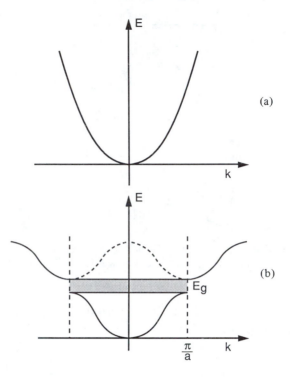

Figure 6.16. (a) Energy vs. wave number curve for a free electron (Parabola).
(b) Energy vs. wave number curve for a nearly free electron in a periodic potential
(first Brillouin zone). The energies in the gap of width E_g are forbidden.

states for the electrons, the gaps corresponding to forbidden states for
which Schrödinger's equation has no acceptable solution. They arise because of the existence of a periodic lattice potential: Electron wave func-
tions represent running waves carrying momentum $\mathbf{p} = \hbar\mathbf{k}$ ($\mathbf{k}$ is the wave
number). At the boundaries of the Brillouin zones the waves cannot prop-
agate (Bragg reflection) and two standing waves are formed with energies
differing by E_g, the gap energy (Fig. 6.16). (We have seen the same effect
for lattice vibrations.)

It is also convenient to envision the formation of bands as due to the
spreading of atomic energy levels (orbitals) when the many atoms constituting the crystal are brought together and the atomic orbitals overlap (Fig.
6.17). The degree of overlap also characterizes the type of atomic bonding.

For ionic bonding, the electrons are transferred from one atomic species
to the other, forming ions with inert gas electronic shell configura-

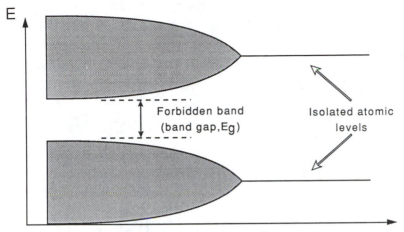

E

Forbidden band
(band gap, Eg)

Isolated atomic
levels

Interatomic Separation

Figure 6.17. Widening of atomic levels into bands when atoms are brought close
together.

tion. There is very little overlap of orbitals and the band formalism
is not adequate. The electrons are bound to the ions and localized.
For covalent bonding, the orbitals overlap and there is some degree of
delocalization of electrons, shared between neighboring atoms.
For metallic bonding, the overlap is large and the electrons belonging to
the outer shell are completely delocalized; there is a sea of nearly
free electrons, bathing the ion core lattice.

The highest occupied band is the *valence band;* the lowest unoccupied
band is the *conduction band.* The gap width is given by

$$E_g = E_c - E_v \qquad (6.105)$$

where E_v is the energy of the top level of the valence band and E_c is the
energy of the bottom level of the conduction band. The electronic levels
are filled up to the *Fermi level E_F.*

The electrical conduction properties of crystals essentially depend on
the position of the Fermi level with respect to the bands and on the gap
width. When an electrical field is applied to a crystal, the electrons are
subjected to a force and are accelerated, thus acquiring energy. They must
therefore be promoted to higher energy levels. This is possible only if the
Fermi level lies within a partially filled conduction band with empty levels
available (metal) or if the gap is narrow enough so that thermal agitation

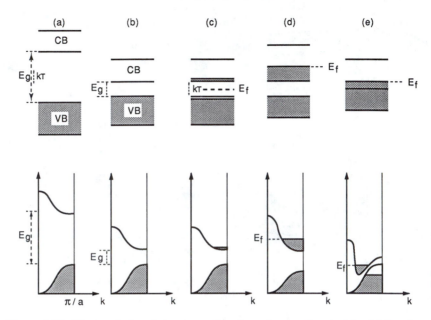

Figure 6.18. Electronic band structure of various classes of materials. (a) Insulators. (b) Intrinsic semiconductors at $T = 0$ K. (c) Intrinsic semiconductors at $T > 0$ K. (d) Metals at $T = 0$ K. (e) Semimetals at $T = 0$ K. The filled levels are stippled.

can allow electrons to cross it, from the valence band into the empty conduction band (semiconductor). If the valence band is completely filled and the gap is wide, the crystal is an insulator, and the electric field cannot make the electrons move since there are no accessible energy levels for them. However, the bands can be distorted, giving rise to polarization (dielectric). We can now classify the crystals according to their electronic structure and electrical properties.

i. Insulators (Fig. 6.18a). The valence band is completely filled at 0 K and the gap is too wide for thermal agitation to allow promotion of electrons across it, even at high temperatures, that is, $E_g \gg kT$.

Typical insulators are, for instance, diamond ($E_g = 5.3$ eV), forsterite ($E_g = 6.4$ eV), and SiO_2 ($E_g = 8$ eV). (Incidentally, remember that $kT = 1/40$ eV at room temperature and that $kT = 1$ eV at 11605 K.) Ionic crystals generally are insulators at low temperatures. This is due to the fact that the charges can only move by thermally activated jumps from one site to another; conduction either involves diffusion of the whole ions (*ionic conduction*) or electronic charge transfer between ions of different valence

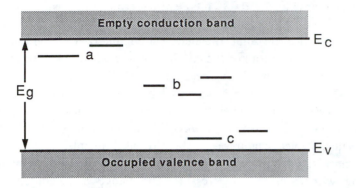

Figure 6.19. Localized levels introduced in the energy gap. (a) Shallow donor
levels. (b) Deep levels. (c) Shallow acceptor levels.

(*hopping conduction*). In both cases, if the activation energy of the process is much higher than kT, the crystal behaves as a dielectric with very
low conductivity. Although the band formalism is not the best suited to
ionic crystals, a large value of the band gap width is associated with ionic
crystals in the low temperature insulating regime. At high temperatures,
ionocovalent compounds that are insulators at low temperatures can simultaneously exhibit electronic semiconduction or hopping conduction
and ionic conduction.

Impurity atoms in low concentrations may introduce deep energy levels
in the gap, far from the gap edges (Fig. 6.19), which may allow for some
electrical conductivity of the semiconduction type if the gap is not too
wide (e.g., Morin, Oliver, and Housley 1977, for forsterite).

ii. Semiconductors (Fig. 6.18b, c). Semiconductors differ from insulators in that their gap is narrower, comparable to kT, thus allowing in pure
crystals thermal promotion of electrons in the conduction band, leaving
positively charged *holes* at the top of the valence band (*intrinsic semiconduction*). Typical gap widths for semiconductors are of the order of
1 eV (e.g., $E_g = 1.14$ eV for Si, $E_g = 0.67$ eV for Ge, and $E_g = 1.4$ eV for
GaAs).

Impurity atoms in solution can introduce shallow levels near the top of
the gap, which can give electrons to the conduction band (*donor levels*)
(e.g., As in Ge), and give rise to *n*-type conduction. If the impurity can
receive electrons, it introduces levels near the bottom of the gap (*acceptor
levels*) (e.g., Ga in Ge), and gives rise to *p*-type conduction, by holes in
the valence band. This is known as *extrinsic semiconduction*.

iii. Metals (Fig. 6.18d). In metals, the Fermi level falls within the conduction band, the remaining empty levels in the band are directly accessible to the electrons, and the electrical conductivity is good at 0 K. In some cases, the bottom of the conduction band is lower than the top of the valence band in certain directions of reciprocal space (*k*-space) and the electrons at the top of the valence band can spill over into the conduction band (Fig. 6.18e). There are conduction electrons and holes at 0 K. Such crystals are known as *semimetals,* which are not as good conductors as metals. Typical examples are the rare-earth metals.

It must be noted that the electrical properties of a given compound can change with temperature or pressure. The change can be gradual and two or more conduction mechanisms may coexist, as when an insulator at low temperatures becomes a semiconductor and/or an ionic conductor at high temperatures or a metallic conductor at high pressure (Samara 1967). The change may also be discontinuous, as when some oxides become metallic conductors above a critical temperature because, for a critical interatomic distance, the electronic structure with localized electrons is less stable than the metallic one (*Mott transition*). Pressure may lower the critical temperature (Mott 1961).

6.5.2 *Mechanisms of electrical conduction*

i. Generalities

Ohm's law, $I = V/R$, relates the current flowing in a conductor to the difference of potential driving it and to the resistance R (or conductance R^{-1}) of the conductor.

The resistivity is $\rho = RS/L$, where S is the cross-sectional area of the conductor and L is its length. Ohm's law can therefore be written

$$\mathbf{J} = \sigma \mathcal{E} \qquad (6.106)$$

where $J = I/S$ (in amperes/m^2) is the current density and $\mathcal{E} = V/L$ (in Volts/m) is the electric field. $\sigma = \rho^{-1}$ (in Ω^{-1}/m or siemens/m) is the electrical conductivity.

The current is equal to the rate of passage of electric charge through a unit area of the conductor; it can be written

$$J = nqv_d \qquad (6.107)$$

where n is the number of mobile charge-carrying particles per unit volume, q is their electric charge, and v_d is their drift velocity in the direction of the electric field.

We will give here a microscopic expression of Ohm's law in the case of electronic conduction in a partially filled band (metal or semiconductor). The electrons feel the periodic potential of the crystal (which accounts for the existence of the bands in the first place), but, for the sake of simplicity and without too much loss of generality, we will consider the conduction electrons as free.

Let us consider electrons moving freely in a band. The force exerted on them by the electric field is

$$\mathbf{F} = -e\mathcal{E} = \frac{d\mathbf{p}}{dt} = \hbar \frac{d\mathbf{k}}{dt} \tag{6.108}$$

The electrons are accelerated by the force but their velocity does not increase indefinitely, because of collisions and scattering of phonons or point defects, and it reaches a steady-state drift velocity v_d. Let the relaxation time τ be the average time between scattering events. The increase in momentum between these events is

$$\hbar \Delta \mathbf{k} = \mathbf{F}\tau \tag{6.109}$$

and the drift velocity is

$$v_d = \hbar \frac{\Delta k}{m} = \frac{F\tau}{m} \tag{6.110}$$

where m is the mass of the electron, or with (6.108),

$$v_d = -e\mathcal{E} \frac{\tau}{m} \tag{6.111}$$

and, with (6.107):

$$J = -nev_d = \frac{ne^2\tau}{m}\mathcal{E}$$

Hence

$$\sigma = \frac{ne^2\tau}{m} \tag{6.112}$$

If Λ is the mean free path of electrons and v is the total average (rms) velocity (thermal velocity plus drift velocity), we have:

$$\sigma = \frac{ne^2\Lambda}{mv} \tag{6.113}$$

If the band is only slightly filled, as in extrinsic n-type semiconductors, the average kinetic energy of an electron is

$$\tfrac{1}{2}mv^2 = \tfrac{3}{2}k_BT \tag{6.114}$$

Hence

$$\sigma = ne^2 \Lambda (3mk_B T)^{-1/2} \tag{6.115}$$

ii. Metallic conduction

All the conduction electrons gain energy when the electric field is applied and most of them are shifted from the levels they occupied to those left vacant by others, with no net contribution to the electrical current. It is only the electrons near the Fermi level that are promoted to empty levels and provide a net response to the field. To first approximation we can take for v in (6.113) the velocity at the Fermi level v_F given by

$$\tfrac{1}{2}mev_F = E_F \tag{6.116}$$

and we obtain *Drude's formula*

▶ $$\sigma = \frac{ne^2 \Lambda}{mv_F} \tag{6.117}$$

The temperature dependence of the conductivity in $T^{-1/2}$ given by the free electron approximation does not in fact agree with experimental results. Indeed, at low temperatures (a few K), the conductivity is controlled by the collision of electrons with dilute impurities in the lattice, depends on the impurity concentration, and is practically independent of temperature. At high temperatures $(T > \Theta_D)$, the conductivity is controlled by phonon scattering and decreases as T^{-1}. The resistivity of metals can be written (*Matthiessen's rule*) as:

$$\sigma^{-1} = \sigma_0^{-1} + \sigma_L^{-1} \tag{6.118}$$

where σ_0^{-1} is the impurity contribution and $\sigma_L^{-1} \propto T$ at high temperatures is the lattice-dependent part. The linear temperature dependence of the resistivity can be qualitatively explained in the following way (Mott and Jones 1958).

The electrical resistance is due to the scattering of electrons by vibrating atoms and it is proportional to the scattering probability, which, in turn, is proportional to the mean square of the displacement of atoms $\langle u^2 \rangle$. At high temperatures $(T > \Theta_D)$, the mean potential energy of vibrating atoms is

$$\tfrac{1}{2}f\langle u^2 \rangle = \tfrac{1}{2}k_B T \tag{6.119}$$

where f is the restoring force. From the equation of motion

$$M\frac{d^2u}{dt^2} + fu = 0 \tag{6.120}$$

we have:

$$f = M\omega^2 \qquad (6.121)$$

where M is the mass of the atoms and ω is their vibrational frequency. With (6.119) and (6.121) and taking for ω the Debye frequency $\omega_D = (k_B/\hbar)\Theta_D$, we obtain

$$\sigma^{-1} \propto \langle u^2 \rangle = \hbar^2 (k_B M)^{-1} \Theta_D^2 T \qquad (6.122)$$

Since $\langle u^2 \rangle$ decreases with pressure, the conductivity of metals increases with pressure. We have, with the definition (3.52) of Grüneisen parameter γ,

$$-\frac{d \ln \sigma}{d \ln V} = \frac{d \ln \sigma}{d \ln \rho} = -2 \frac{d \ln \Theta_D}{d \ln V} = 2\gamma \qquad (6.123)$$

iii. Semiconduction

In the case of semiconductors, where electron or hole states are close to the edge of a band, the free electron approximation cannot be used and the interaction of the electrons with the lattice and other electrons must be taken into account by introducing the *effective mass m**, instead of the mass m of the free electron. The expression of the effective mass can be found by considering the motion of a wave packet of states with frequencies and wave-numbers near ω and $\mathbf{k}$, respectively. Energy $E = \hbar\omega$ is propagated with group velocity v_g:

$$v_g = \frac{d\omega}{dk} = \frac{1}{\hbar} \frac{dE}{dk} \qquad (6.124)$$

$$\frac{dv_g}{dt} = \frac{1}{\hbar} \frac{d^2E}{dk^2} \frac{dk}{dt}$$

With $\hbar d\mathbf{k}/dt = \mathbf{F}$ (6.108), it becomes

$$\mathbf{F} = \hbar \left(\frac{d^2E}{dk^2} \right)^{-1} \frac{dv_g}{dt} = \frac{d\mathbf{p}}{dt} \qquad (6.125)$$

We see that the role of the mass is played by the effective mass

$$m^* = \hbar^2 \left(\frac{d^2E}{dk^2} \right)^{-1} \qquad (6.126)$$

For free electrons $E = \hbar^2 k^2/2m$, hence $m^* = m$. Close to the edge of the bands, Bragg reflection, representing momentum transfer between the electrons and the lattice, is thus taken into account by introducing the effective mass.

For a semiconductor with band gap width E_g and Fermi level E_F, if the energy at the top of the valence band is taken as the zero of energy, the number of electrons in the conduction band (filled states) per unit volume is

$$n = \int_{E_g}^{\infty} 2\mathfrak{D}(E) f(E) \, dE \tag{6.127}$$

where $\mathfrak{D}(E)$ is the density of states of electrons (the number of states per unit volume with energy between E and $E + dE$) and $f(E)$ is the Fermi–Dirac distribution function:

$$f(E) = \left[\exp\left(\frac{E - E_f}{k_B T} \right) + 1 \right]^{-1} \cong \exp\left(\frac{E_F - E}{k_B T} \right) \tag{6.128}$$

Calculating the value of the density of states (see, e.g., Kittel 1967; Animalu 1977), it can be shown that

$$n = 2 \left(\frac{m_e^* k_B T}{2\pi \hbar^2} \right)^{3/2} \exp\left(\frac{E_F - E_g}{k_B T} \right) \tag{6.129}$$

Similarly, the concentration of holes in the valence band is

$$p = 2 \left(\frac{m_h^* k_B T}{2\pi \hbar^2} \right)^{3/2} \exp\left(\frac{-E_F}{k_B T} \right) \tag{6.130}$$

The law of mass action for the concentrations of electrons and holes is written as:

$$np = 4 \left(\frac{k_B T}{2\pi \hbar^2} \right)^3 (m_e^* m_h^*)^{3/2} \exp\left(\frac{-E_g}{k_B T} \right) \tag{6.131}$$

where m_e^* and m_h^* are the effective masses of electrons and holes, respectively.

i. Intrinsic conduction. For intrinsic conduction, we have $n = p$, since a hole is left in the valence band for every electron that is promoted to the conduction band. The number of intrinsic carriers is then found from (6.131) to be

$$n_i = p_i = 2 \left(\frac{k_B T}{2\pi \hbar^2} \right)^{3/2} (m_e^* m_h^*)^{3/4} \exp\left(\frac{-E_g}{2k_B T} \right) \tag{6.132}$$

and, setting (6.129) equal to (6.130),

$$E_F = \frac{1}{2} E_g + \frac{3}{4} k_B T \ln\left(\frac{m_h^*}{m_e^*} \right)$$

One then sees that, if $m_e^* = m_h^*$, the Fermi level lies in the middle of the gap. The conductivity is given by the sum of the contributions of electrons and holes:

$$\sigma = ne\mu_e + pe\mu_h = n_i e(\mu_e + \mu_h) \tag{6.133}$$

where μ_e and μ_h are the *mobility* of electrons and holes, respectively (drift velocity per unit electric field). Comparing (6.133) with (6.112) and replacing m by the effective masses yields:

$$\mu_e = e\tau_e m_e^*$$
$$\mu_h = e\tau_h m_h^* \tag{6.134}$$

where τ_e and τ_h are the relaxation times for electrons and holes, respectively.

From (6.132) and (6.133), we have therefore for intrinsic conductivity:

$$\sigma = 2e\left(\frac{k_B T}{2\pi\hbar^2}\right)^{3/2} (m_e^* m_h^*)^{3/4} \exp\left(\frac{-E_g}{k_B T}\right)(\mu_e + \mu_h) \tag{6.135}$$

Intrinsic conductivity is the conduction regime for pure crystals or crystals with low impurity concentrations at high temperature, where the concentration of intrinsic carriers, which grows exponentially with temperature, is much larger than the concentration of impurities.

The mobility at high temperatures is mostly controlled by phonon scattering and it can be shown that

$$\mu_e = \mu_h \propto T^{-3/2} \tag{6.136}$$

The temperature dependence of the conductivity therefore comes essentially from the temperature dependence of the carrier concentration. The slope of the Arrhenius plot directly gives the energy gap E_g (Fig. 6.20).

ii. Extrinsic conduction. The electrons in the conduction band of *n*-type semiconductors (holes in the valence band of *p*-type semiconductors) are yielded by thermal ionization of the donor (acceptor) impurities. The donor level (E_d below the conduction band) is at $E_g - E_d$, with respect to the top of the valence band. The conduction electron concentration n is equal to the concentration of ionized donors N_d^+, given by the product of the total dopant concentration by the Fermi–Dirac distribution:

$$n = 2\left(\frac{m_e^* k_B T}{2\pi\hbar^2}\right)^{3/2} \exp\left(\frac{E_F - E_g}{k_B T}\right) = N_d\left[\exp\left(\frac{E_F - E_g + E_d}{k_B T}\right) + 1\right] \tag{6.137}$$

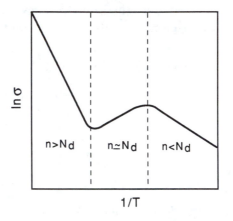

Figure 6.20. Arrhenius plot of the electrical conductivity of semiconductor showing the intrinsic regime at high temperatures and the extrinsic regime at low temperatures; in the intermediate regime the conductivity may slightly decrease with increasing temperature due to electron–phonon scattering (after Animalu 1977).

In the low temperature limit, $E_d \gg k_B T$, we have:

$$\exp 2\left(\frac{E_F - E_d}{k_B T}\right) \approx \frac{1}{2} N_d \left(\frac{m_e^* k_B T}{2\pi \hbar^2}\right)^{-3/2} \exp\left(\frac{-E_d}{k_B T}\right)$$

and

$$n \approx (2N_d)^{1/2} \left(\frac{m_e^* k_B T}{2\pi \hbar^2}\right)^{3/2} \exp\left(\frac{-E_d}{2k_B T}\right) \tag{6.138}$$

The carrier concentration varies with the square root of the impurity concentration and as $T^{3/2}$. The mobility, controlled by ionized impurity scattering, varies as $T^{-3/2}$. The temperature dependence of the conductivity is therefore exponential, with the slope of the Arrhenius plot giving the ionization energy of the impurity E_d (Fig. 6.20). The same results apply to p-type semiconductors.

Between the intrinsic and the extrinsic regimes, there is an intermediate temperature range where the concentration of intrinsic carriers is comparable with that of the dopant impurity. The conductivity in this range may slightly decrease as temperature increases because of electron–phonon scattering (Fig. 6.20).

Pressure affects the electrical conductivity of semiconductors via the mobility and concentration of the carriers. Mobility controlled by phonon scattering depends on, for example, density, phonon velocity, and effective mass, all of which depend on pressure (Paul and Warschauer 1963). The overall pressure dependence of mobility, however, is of second order compared to the pressure dependence of the carrier concentration: The carrier concentration depends exponentially on the energy gap E_g, which depends in a complicated way on pressure. At pressures up to 10 kbar, for instance, the energy gap of germanium first increases with pressure (5×10^{-6} eV/bar), then tends to decrease, whereas the energy gap of silicon decreases as pressure increases (Paul and Warschauer 1963). At very high pressures, anyway, one should expect that bands would eventually overlap, giving rise to metallic conduction (Drickamer 1963).

iii. Hopping conduction and ionic conduction. Let us now consider the case of ionic crystals where the orbital overlap is so small that the band formalism becomes inadequate. To first approximation, the electrons are localized and bound to the ions, they cannot freely move as an electric field is applied, and the crystal behaves as an insulator at low temperatures. Such compounds, for which the effective mass of the carriers is very large, are sometimes called "low-mobility semiconductors." The only way an electrical current can be created is by bodily motion of the ions by thermally activated diffusion through the crystal (ionic conduction) or by thermally activated charge transfer between neighboring ions of different valence (hopping conduction): for instance, a hole hops from a cation with charge $+3$ to a neighboring ion with charge $+2$ or $+1$. This occurs in the so-called mixed-valence compounds. The ions can be of different natures (e.g., Ni^{3+} and Li^{+} in lithium-doped NiO). They can also be of the same nature, as in the important case of transition metals, which can exist with several degrees of oxidation, for example, Fe^{3+} and Fe^{2+} in magnetite Fe_3O_4 (Kündig and Hargrove 1969); see also Coey et al. (1989).

There are also cases where the low-mobility carriers in narrow-band materials, interacting strongly with optical phonons, polarize the lattice and distort it in their neighborhood. The lattice distortion moves with the carrier (see Appel 1968). The unit formed by the bound carrier and its induced lattice deformation, confined to a small region, is called a *small polaron*. The strain energy of the distorted lattice E_b is equal to the polaron binding energy. Conduction by polarons has been found in the case of oxides such as NiO, CeO_2, and FeO (Yamashita and Kurosawa 1958; Austin and Mott 1969; Tuller and Nowick 1977; Chen, Gartstein, and Mason 1982). At high temperatures, small polarons move by thermally activated hopping (Emin 1975). We will limit ourselves to the discussion of hopping and ionic conduction, both of which proceed by thermally activated diffusion jumps.

The *Nernst–Einstein relation* links the electrical conductivity to the diffusion coefficient of electrons or ions; it is obtained by equating the expression of Ohm's law (6.106) with the diffusion flux of charges:

$$\sigma \mathcal{E} = -D \operatorname{grad}(nq) = -qD \frac{\partial n}{\partial x} \tag{6.139}$$

where q is the charge of the carrier (ion, electron or hole) diffusing under the action of the electric field $\mathcal{E}$. When a steady state is reached, the concentration $n(x)$ of carriers at point x in the electric field follows a Boltzmann distribution law:

$$n(x) \propto \exp\left(\frac{-e\mathcal{E}x}{k_B T}\right) \tag{6.140}$$

From (6.139) and (6.140), we have:

$$\frac{\sigma}{D} = \frac{q^2 n}{k_B T} \tag{6.141}$$

For the case of hopping, the electron (hole) transfer reaction can be written, for instance:

$$M_a^{3+} + M_b^{2+} \to M_a^{2+} + M_b^{3+} \tag{6.142}$$

where M_a^{q+} and M_b^{q+} are cations of charge q in neighboring sites a and b.

If θ is the fraction of lattice sites occupied by electrons or holes capable of jumping (e.g., atomic fraction of M^{3+}), the probability of finding an empty site in the neighborhood of an electron or a hole is $1 - \theta$ and the diffusion coefficient of charges (see Honig 1970) is

$$D = \Gamma(\delta l)^2 (1 - \theta) \tag{6.143}$$

where Γ is the jump frequency and δl is the jump distance (see Sec. 6.2). The jump frequency is thermally activated:

$$\Gamma = \Gamma_0 \exp\left(\frac{-\Delta H}{k_B T}\right) \tag{6.144}$$

The number of carriers per unit volume n is proportional to θ, the constant of proportionality depending on the crystal structure. With (6.143) and (6.144), the Nernst–Einstein relation gives:

$$\sigma = \frac{\sigma_0 \theta(1 - \theta)}{k_B T} \exp\left(\frac{-\Delta H}{k_B T}\right) \tag{6.145}$$

where σ_0 is a constant and ΔH is the activation energy for hopping. If the atomic fraction θ of charge carriers is constant, the hoping conductivity depends on temperature according to an Arrhenius law like diffusion and it is a maximum for $\theta = 0.5$. Deviations from an Arrhenius law obviously occur if θ changes with temperature, for example, if the concentration of Fe^{+3} increases due to oxidation of Fe^{+2} as the oxygen partial pressure increases. The case of ionic conduction obviously reduces, via the Nernst–Einstein relation, to the case of ionic diffusion, treated in Section 6.2.

The effect of pressure, however, is quite different for hopping and ionic conductivity. In the case of ionic conductivity, we have seen in Section 6.2 that the activation energy for diffusivity increases with pressure because it is more difficult to create vacancies under pressure. The ionic conductivity therefore decreases as pressure increases. On the contrary, the activation energy for hopping conductivity decreases as pressure increases because the jump distance between sites decreases under pressure. As there is no need for vacancies, the hopping conductivity increases with pressure.

6.5.3 Electrical conductivity of mantle minerals

A review of the experimental measurements of the conductivity of various minerals and rocks at high pressure was given by Parkhomenko (1982). We will only briefly summarize here the results concerning the main mantle minerals.

i. Forsterite, olivines, fayalite

Pure forsterite Mg_2SiO_4 is an insulator at room temperatures ($\sigma < 10^{-7}$ S/m). Its conductivity was studied by Bradley, Jamil, and Munro (1964), who found that its activation energy between 300 and 700°C increases with pressure, from 1.1 eV at 11 kbar to about 3 eV at 35 kbar. Measurements on single crystals in the three directions a, b, and c (Morin, Oliver, and Housley 1977, 1979) showed a conductivity anisotropy, with the highest conductivity in the a direction and the lowest conductivity in the b direction. These results are not confirmed by the recent measurements between 1000 and 1500°C by Schock, Duba, and Shankland (1989), who find that the conductivity is greatest and independent of oxygen partial pressure in the c direction, intermediate in the a direction, and least in the b direction; they infer that conduction in forsterite is dominated by electronic conduction in the a and b directions and probably by magnesium vacancy diffusion along the c direction. Will et al. (1979) studied

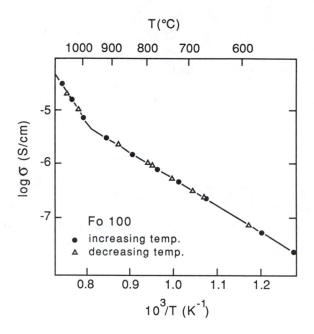

Figure 6.21. Arrhenius plot of the electrical conductivity of forsterite equilibrated with MgO (after Cemic et al. 1980).

the electrical conductivity of forsterite equilibrated with MgO, the thermodynamic conditions being thus well defined, they found two regimes (Fig. 6.21):

$$\sigma = 4.92 \exp\left(\frac{-0.98 \text{ eV}}{k_B T}\right) \quad \text{for } 520 < T < 970°C$$

$$\sigma = 4.69 \times 10^{-6} \exp\left(\frac{-2.46 \text{ eV}}{k_B T}\right) \quad \text{for } 970 < T < 1075°C$$

Olivines $(Mg_{1-x}Fe_x)_2SiO_4$ and fayalite Fe_2SiO_4 were studied by many authors. Schock et al. find that olivine appears to show mixed conduction from 1000 to 1500°C, as evidenced by the curvature of the Arrhenius plot (Fig. 6.22), with electron hole conduction by hopping predominant at lower temperatures and magnesium vacancy ionic conduction predominant above 1390°C.

Even though the results are somewhat scattered (Table 6.3), most investigations lead to the same conclusions:

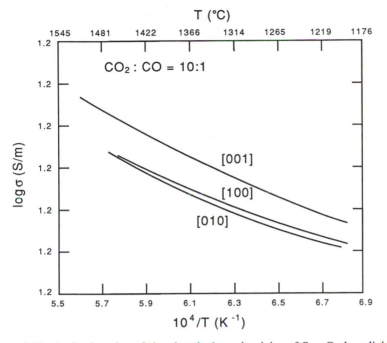

Figure 6.22. Arrhenius plot of the electrical conductivity of San Carlos olivine. The gas mixture (10 CO_2 for 1 CO) buffers the oxygen partial pressure at the value of 10^{-3} Pa at 1200°C (after Schock et al. 1989).

The conductivity of olivines increases and the activation energy decreases as the total iron content x increases (Bradley, Jamil, and Munro 1964; Cemic, Will, and Hinze 1980).

The conductivity increases and the activation energy decreases as the ratio of ferric to ferrous iron content increases, for instance, when measurements are made in an oxidizing atmosphere (Duba, Ito, and Jamieson 1973; Duba and Nicholls 1973; Cemic et al. 1980).

The conductivity increases with pressure; its activation energy is not seen to significantly vary with pressure up to 8 kbar (Duba, Heard, and Schock 1974) but linearly decreases as pressure increases from 10 to 60 kbar, with $dE/dP = -5.2$ eV/kbar (Bradley et al. 1964). Shockwave experiments up to 560 kbar (Mashimo et al. 1980) confirm these results.

The same conclusions apply to other orthosilicates of transition metals, such as manganese, cobalt, and nickel olivines (Bradley, Milnes, and

Table 6.3. *Parameters of the Arrhenius law for olivines*
$(Mg_{1-x}Fe_x)_2SiO_4$

x (% Fe)	T range (°C)	P (kbar)	σ_0 (S/m)	E (eV)	Reference
10.4–17.5	560–1800	11.5–32		0.9	Hamilton (1965)
1	300–1000	12	11200	1	Bradley et al. (1964)
"	" "	35	230	1	"
10	" "	12	200	0.92	"
"	" "	47	5	0.74	"
50	" "	12	17380	0.74	"
100	" "	12	1900	0.72	"
"	" "	47	370	0.53	"
18	500–1200	0	29	0.86	Mizutani and Kanamori (1967)
8.4	800–1200	0	160	0.79	Kobayashi and Maruyama (1971)
19[a]	560–1120	2–8	1.3	0.98	Duba et al. (1974)
"	1270–1440	"	53700	2.33	"
10[b]	870–1210	10	2.7	0.78	Cemic et al. (1980)
10[c]	820–1220	10	1.9	0.62	"
20[b]	800–1210	10	13.8	0.68	"
20[c]	875–1160	10	16.3	0.58	"
100[d]	340–1100	10		0.52	Will et al. (1979)
100[e]	" "	10		0.38	"
9[f]	1200–1250	0		1.3	Schock et al. (1989)
9[g]	1200–1250	0		1.6	"

Note: $\sigma = \sigma_0 \exp(-E/k_B T)$.
[a] Red Sea olivine with a low Fe^{3+}/Fe^{2+} ratio
[b] Reducing conditions (Fe–FeO buffer)
[c] Oxidizing conditions (Quartz–Magnetite–Fayalite buffer)
[d] Reducing conditions (Quartz–Fe)
[e] Oxidizing conditions (Quartz–Magnetite)
[f] $f_{O_2} = 10^{-4}$ Pa, a and b directions
[g] $f_{O_2} = 10^{-4}$ Pa, c direction

Munro 1973), and to manganese germanate olivine (Yagi and Akimoto 1974).

The experimental results for lower temperatures up to about 1200°C are generally consistent with a conduction mechanism by electron transfer by hopping between Fe^{3+} and Fe^{2+} ions (Bradley et al. 1964, 1973).

The transition from olivine (fayalite) to spinel is accompanied by an increase in conductivity by about two orders of magnitude at 900 K (Akimoto and Fujisawa 1965). This is consistent with the hopping mechanism, which is easier in the spinel structure (as, for instance, in magnetite).

ii. Periclase and magnesiowüstite

Mitoff (1962) measured the conductivity of single crystals of MgO at high temperature and various oxygen partial pressures. He found that the conduction was predominantly ionic at 1000°C and intermediate oxygen partial pressures, whereas at 1500°C and high (1 atm) or low ($<10^{-6}$ atm) pressures, the conduction was mostly electronic, with deviation from stoichiometry resulting in sources of electronic charge carriers. Periclase is, in most cases, at high temperatures, a mixed electronic and ionic conductor.

The behavior of the electrical conductivity of magnesiowüstites $(Mg_{1-x}Fe_x)O$ and wüstite $Fe_{1-x}O$, as a function of pressure, total Fe content, and ferric/ferrous iron ratio, is very similar to that of the olivines. It has also generally been ascribed to hopping (Tannhauser 1962; Hansen and Cutler 1966; Iyengar and Alcock 1970; Mao 1972; Bowen, Adler, and Auker 1975; Chen et al. 1982).

The conductivity of magnesiowüstites and wüstite is much higher than that of olivines and its activation energy is lower, ranging from 0.7 to 0.4 eV for iron contents of 7.5 to 20 percent at ambient pressure (Iyengar and Alcock 1970). Mao (1972) found that at 150 kbar, the electrical conductivity of $(Mg_{0.78}Fe_{0.22})O$ is six orders of magnitude higher than that of olivines with the same Fe/Mg ratio, with an activation energy equal to 0.37 eV.

Shock-wave experiments at pressures above 700 kbar and temperatures above 1000 K have shown that wüstite $Fe_{0.94}O$ undergoes a transition to the metallic state (Knittle and Jeanloz 1986).

iii. Crystals with perovskite structure

Silicate perovskite $(Mg, Fe)SiO_3$ is thought to be an essential constituent of the lower mantle, but so far, very few experiments on its electrical conductivity have been reported in the literature and they are conflicting: Li and Jeanloz (1987) report on the results of measurements in the laser-heated diamond-anvil cell up to 850 kbar and claim that the conductivity is lower than 0.01 S/m up to the melting point, almost independently of temperature; Peyronneau and Poirier (1989), however, measured the conductivity of the perovskite at 400 kbar, in an externally heated diamond-anvil cell up to about 400°C, and they found that the conductivity is already 0.01 S/m at 400°C and that it increases with temperature (activation energy 0.35 eV), pressure, and iron content in a way entirely consistent with a hopping mechanism (Fig. 6.23).

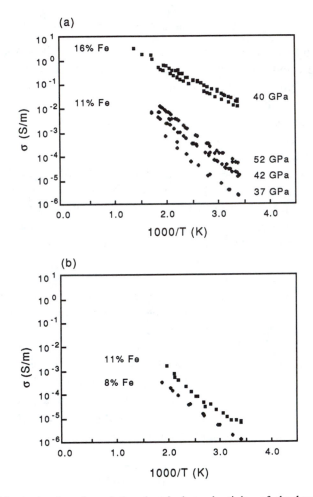

Figure 6.23. Arrhenius plot of the electrical conductivity of the lower mantle material. (a) Perovskite and magnesiowüstite assemblage obtained by decomposition of San Carlos olivine of various Fe contents. (b) Perovskite $(Mg, Fe)SiO_3$ with 8 and 11% Fe at 42 GPa. (After Peyronneau and Poirier 1989)

Some other perovskites have been investigated and are generally thought to be electronic extrinsic semiconductors up to the highest temperatures investigated (generally about 1000°C). This is in particular the case for $SrTiO_3$, with an activation energy of 0.4 eV above 1000 K (Stumpe, Wagner, and Bäuerle 1983). The conductivity of $CaTiO_3$ has also been investigated up to 1100°C. It exhibits several regimes of conduction depending on the oxygen partial pressure, which are interpreted in terms of small amounts of acceptor impurities (Balachandran, Odekirk, and Eror 1982).

Fluoride perovskites $NaMgF_3$ (O'Keefe and Bovin 1979) and $KZnF_3$ (Poirier et al. 1983) were reported to exhibit a solid-electrolyte-like behavior close to the melting point, attributed to an abnormally high mobility of quasi-delocalized fluorine anions, which gives a very high electrical conductivity comparable to that of a liquid. Kapusta and Guillopé (1988) and Wall and Price (1989) performed molecular dynamics simulations on $MgSiO_3$ and found a very high mobility of the oxygen ions, which lead them to the conclusion that the silicate perovskite might also behave as a solid electrolyte at high temperatures. There is, however, no experimental support so far for this hypothesis.

6.5.4 Electrical conductivity of the core

Secco and Schloessin (1989) have recently measured the electrical conductivity of pure iron in the solid and liquid state as a function of temperature and at pressures up to 7 GPa. Extrapolating their results to the P, T conditions of the outer core, they find it improbable that the conductivity of the impure outer core can differ much from that of pure liquid iron at pressures below 7 GPa and conclude that the most probable range of its value is $0.67–0.83 \times 10^6$ S/m.

6.6 Thermal conduction

i. Generalities

Heat is transferred through solids by essentially two processes: lattice conduction (or phonon conduction) and electronic conduction. In the former case, the carriers are quantized lattice vibrations, that is, phonons, and in the latter case, electrons. Both processes operate in some measure in all solids, but it is obvious that, in electrical insulators, phonons are responsible for almost all of the conductivity, whereas in metals, electronic conduction plays a predominant role. In both cases, the principal features of the mechanisms of conduction can be accounted for by the classical (or semiclassical) approach of the kinetic theory of gases, considering a gas of electrons or phonons. We will limit ourselves here to this elementary outlook, referring the reader to the book by Berman (1976) for a more rigorous treatment. Techniques of measurement are dealt with in Horai and Shankland (1987) and values of the thermal conductivities of many rock-forming minerals can be found in Horai (1970).

The geophysical literature usually lumps together thermal conduction and radiative heat transfer. This is somewhat misleading since thermal conduction refers to the transport of heat, the degraded form of energy, whereas radiative transfer concerns the propagation of energy, carried by electromagnetic waves (photons) through a more or less transparent (i.e., nonabsorbing) medium. A justification can be found in the fact that at high temperatures, most experimental methods for measuring the thermal conductivity of solids yield in fact an effective thermal conductivity, comprising a term corresponding to radiative transfer (Kanamori, Fujii, and Mizutani 1968). We will follow the common usage and briefly deal with radiative transfer at the end of this section.

ii. Lattice conduction

The carriers of heat are phonons. In an anharmonic crystal, phonons interact (collide) and the thermal conductivity is controlled by the mean free path Λ between collisions: The more collisions there are, the shorter the mean free path and the smaller the conductivity. In a harmonic crystal, with no interactions among phonons, the mean free path, hence the conductivity, would be infinite.

Let us consider thermal conduction in a material along the x direction. The heat flux is given by

$$J = -n(c\Lambda \operatorname{grad} T)v_x \tag{6.146}$$

where n is the number of carriers per unit volume, c is their specific heat ($c\Lambda \operatorname{grad} T$ is the amount of heat transferred over the length of the mean free path), and v_x is their average velocity in the x direction. We can also write

$$J = -\tfrac{1}{3}Cv\Lambda \operatorname{grad} T \tag{6.147}$$

where $C = cn$ is the lattice specific heat per unit volume and $v = 3v_x$ is the root mean square velocity of the phonon gas, that is, the sound velocity. Comparison with equation (6.20) gives the lattice heat conductivity

$$k_1 = \tfrac{1}{3}Cv\Lambda \tag{6.148}$$

At low temperatures ($T < \Theta_D$) the lattice specific heat varies as T^3 and the mean free path Λ tends to become constant and controlled by defects, hence the thermal conductivity varies as T^3 (Fig. 6.24). At high temperatures ($T > \Theta_D$) the specific heat tends toward $3R$ (R is the gas constant) and it can be more or less empirically shown (Dugdale and McDonald 1955) that the mean free path varies as $1/T$, and thermal conductivity

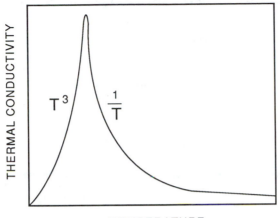

Figure 6.24. Temperature dependence of the lattice thermal conductivity.

therefore decreases as $1/T$ at high temperatures (Fig. 6.24). Minerals of geophysical interest should not depart markedly from this behavior (Roufosse and Klemens 1974).

Since the lattice heat conductivity is linked to anharmonicity, it clearly must be possible to express it in terms of the Grüneisen parameter. Let us start with equation (6.148) and with the expression of Dugdale and McDonald (1955) for the mean free path:

$$\Lambda = \frac{a}{\alpha \gamma T} \tag{6.149}$$

where a is an interatomic distance, α is the thermal expansion coefficient, and γ is the Grüneisen parameter. Taking for γ the thermodynamic gamma

$$\gamma = \frac{\alpha K}{C} \tag{6.150}$$

where $C = \rho C_p$ is the specific heat per unit volume and K is the bulk modulus, we obtain

$$k_1 = \frac{a v K}{3 \gamma^2 T} \tag{6.151}$$

Taking for the average phonon velocity $v = v_\phi = (K/\rho)^{1/2}$, we can write

$$k_1 = \frac{a v^3 \rho}{3 \gamma^2 T} \tag{6.152}$$

and obtain the expression first given by Lawson (1957):

$$k_1 = \frac{aK^{3/2}}{3\gamma^2\rho^{1/2}T} \tag{6.153}$$

The thermal conductivity can also be expressed in terms of the Debye temperature Θ_D. Assuming the minimum wavelength of the vibrational modes to be equal to $2a$, the maximum frequency is

$$\omega_{max} \approx \frac{\pi\hbar v}{a} = \frac{k_B\Theta_D}{\hbar} \tag{6.154}$$

Replacing in (6.151) v by its expression in terms of the Debye temperature and writing

$$K \propto v^2 M a^{-3}$$

where M is the average atomic weight, we obtain (Berman 1976)

$$k_1 \propto \frac{Ma\Theta_D^3}{\gamma^2 T} \tag{6.155}$$

Horai and Simmons (1969, 1970) found empirical relations among the thermal conductivity and the compressional and shear wave velocities:

$$v_P = 0.17k_1 + 5.93$$
$$v_S = 0.09k_1 + 3.31 \tag{6.156}$$

and between thermal conductivity and Debye temperature:

$$\Theta_D = 25.6k_1 + 385 \tag{6.157}$$

with k_1 in mcal/cm s deg, v_P and v_S in km/s, and Θ_D in K (Fig. 6.25).

Using D. L. Anderson (1967) seismic equation of state and introducing the concept of mean atomic specific heat, Maj (1978) derived a relation between the lattice thermal conductivity and the seismic parameter for silicate minerals:

$$k_1 = 0.43\Phi^{0.823} \tag{6.158}$$

where k_1 is expressed in mcal/cm s K and Φ in km^{2}s^{-2}.

An order of magnitude of the thermal conductivity of insulators can be found (Animalu 1977) by setting in (6.148) $\Lambda = 3\times 10^{-6}$ cm, $v = 10^5$ cm/s, $C = 3R$ cal/mol. One obtains

$$k_1 \approx \frac{0.3R}{V_{mol}}$$

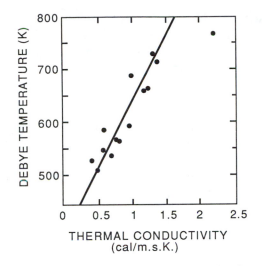

Figure 6.25. Correlation of Debye temperature and lattice thermal conductivity for silicates (after Horai and Simmons 1970).

with typical values of the molar volume V_{mol} of a few tens of cubic centimeters, conductivity values are found to lie in the range of about 5 to 30 mcal/cm s deg (0.8 to 12.5 W/m deg).

The effect of pressure on thermal conductivity can be found through the theoretical or empirical relations between k_1 and the sound velocity: k_1 obviously increases with pressure. For typical mantle minerals, thermal conductivity can undergo a tenfold increase between ambient conditions and the pressure of the core–mantle boundary (Mao 1972). From equation (6.153) we can also see that, along an adiabatic temperature gradient, we have:

$$\frac{d \ln k_1}{d \ln \rho} = -\frac{1}{3} + \frac{3}{2}\frac{d \ln K}{d \ln \rho} - \frac{1}{2}\frac{d \ln T}{d \ln \rho} - \frac{2d \ln \gamma}{d \ln \rho}$$

or, using $(d \ln T/d \ln \rho)_{ad} = \gamma$ and $\gamma\rho = \text{const.}$,

$$\blacktriangleright \qquad \frac{d \ln k_1}{d \ln \rho} = 2\gamma + \frac{5}{3} \qquad (6.159)$$

Using interatomic potentials, Roufosse and Jeanloz (1983) predicted the effect of phase transitions on the thermal conductivity of alkali halides; they suggested that the increase in coordination associated with a small volume change acts to reduce the thermal conductivity at high pressures.

Table 6.4. *Thermal diffusivity of some minerals*

Mineral	T (K)	P (kbar)	κ ($10^{-2}\,cm^2s^{-1}$)	Reference
Quartz ∥ [001]	1100	0.001	1.64	Kanamori et al. (1968)
Quartz ∥ [010]	1100	0.001	1.28	"
Olivine ∥ [001]	1100	0.001	1.35	"
Forsterite	1100	30	0.94	Fujisawa et al. (1968)
	1100	50	1.10	"
NaCl	1000	29	1.20	"
Periclase	1100	0.001	3.23	Kanamori et al. (1968)
Jadeite	1100	0.001	0.96	"
Garnet	1100	0.001	0.83	"
Spinel	1100	0.001	2.13	"

The thermal diffusivity $\kappa = k_1/\rho C_p$ of rock-forming minerals at high temperatures and pressures has been measured by Kanamori et al. (1968) and Fujisawa et al. (1968) (Table 6.4).

We see that for most minerals (and rocks) κ is of order $10^{-2}\,cm^2s^{-1}$. The pressure derivative $d\kappa/dP$ at 40 kbar for Mg_2SiO_4 has been found to be $1.8 \times 10^{-4}\,cm^2s^{-1}kbar^{-1}$ at 700 K and $0.8 \times 10^{-4}\,cm^2s^{-1}kbar^{-1}$ at 1100 K (Fujisawa et al. 1968). Using the volume dependence of the Grüneisen parameter, Kieffer (1976) built a model for the lattice thermal conductivity of the mantle; she found it to be minimal at the depth of the olivine–spinel transition and reach the value $k \approx 0.01$ cal/cm s at the core–mantle boundary.

iii. Electronic conduction

The carriers are electrons instead of phonons but the same general formula (6.148) applies:

$$k_e = \tfrac{1}{3}Cv_F\Lambda \qquad (6.160)$$

where k_e is the electronic conductivity, $C = nc$ is the electronic specific heat per unit volume (n is the number of electrons per unit volume and c is the specific heat of an electron), v_F is the velocity of electrons at the Fermi level, and Λ is their mean free path between collisions. The electrons in a metal are also the carriers of electric current and the electrical conductivity σ was given in the free electron approximation by the Drude formula (6.117):

$$\sigma = \frac{ne^2\Lambda}{m_e v_F} \qquad (6.161)$$

We therefore have

$$\frac{k_e}{\sigma} = \frac{1}{3}\frac{m_e v_F^2 c}{e^2} \qquad (6.162)$$

With $c = \frac{3}{2}k_B$ and $\frac{1}{2}m_e v_F^2 = \frac{3}{2}k_B T$, we have

$$\frac{k_e}{\sigma T} = \frac{3}{2}\left(\frac{k_B}{e}\right)^2 \qquad (6.163)$$

A rigorous quantum mechanical treatment (Ziman 1965) would give

$$\blacktriangleright \qquad \frac{k_e}{\sigma T} = \frac{\pi^2}{3}\left(\frac{k_B}{e}\right)^2 = L_0 = 2.45\times10^{-8}\,\mathrm{W\Omega\,deg}^{-2} \qquad (6.164)$$

This is the *Wiedemann–Franz* law. L_0 is known as the Lorentz number.

The Wiedemann–Franz law is generally well verified for metals, with values of the ratio in rather good agreement with the theoretical value of L_0 (2.47×10^{-8} for Fe and 2.23×10^{-8} for Cu at 0°C).

At low temperatures, the mean free path of electrons is constant and determined by the distribution of defects and impurities, the electrical conductivity is constant, and, by the Wiedemann–Franz law, we see that the thermal conductivity varies linearly with T.

At high temperatures, the electrons are scattered by phonons and their mean free path is proportional to $1/T$, hence, the electronic thermal conductivity varies as $1/T$.

The order of magnitude of k_e can be calculated (Animalu 1977), taking for typical values $v_F \approx 10^8$ cm/s, $\Lambda \approx 10^{-5}$ cm, and $C \approx 0.1R$ cal/mol. It is found to be two orders of magnitude larger than for lattice conductivity:

$$k_e \approx \frac{30R}{V_{\mathrm{mol}}}$$

Metals are normally much better heat conductors than insulators; however, if the maximum of lattice conductivity occurs at relatively high temperature, the heat conductivity of an electrical insulator may be quite large. This is the case for diamond, whose heat conductivity at room temperature is about 2000 W/mK, compared to 400 W/mK for copper.

iv. Radiative conductivity

In nonopaque media, a sizable fraction of energy transfer at high temperatures may occur by thermal radiation (photon transfer). The *opacity* ϵ of a medium is defined by the decrease of the intensity, due to absorption and scattering, of a pencil of radiation passing through a thickness x of material:

$$I = I_0 \exp(-\epsilon x) \qquad (6.165)$$

The opacity is the reciprocal of the mean free path of radiation.

In the simplest case, in which the opacity is assumed to be independent of the wavelength of the radiation, the *radiative conductivity* k_r can be written (Clark 1957):

$$k_r = \frac{16 n^2 S T^3}{3\epsilon} \qquad (6.166)$$

where n is the index of refraction and S is the Stefan–Boltzmann constant. With $n \approx 1.7$ (a typical value for ferromagnesian silicates), one gets

$$k_r \approx 9.2 \times 10^{-9} \frac{T^3}{\epsilon} \ \text{W/mK} \qquad (6.167)$$

One of the important mechanisms responsible for opacity in the ferromagnesian silicates is the absorption of photons causing the charge transfer from Fe^{2+} to Fe^{3+} (or equivalently, the excitation of electrons from one narrow band to another). We have seen that the same charge transfer process can be thermally activated and is then responsible for hopping electrical conduction. High pressure, favoring overlapping of the electronic orbitals of neighboring ions, increases the optical absorption due to charge transfer and causes the absorption edge to shift from the ultraviolet to the visible and infrared regions of the spectrum. It is then possible for energy transfer by radiation in the lower mantle to be effectively blocked, the increase of opacity with pressure overriding the T^3 dependence of the radiative conductivity (Mao 1972, 1976).

7

Earth models

Inferences about the interior of the Earth, so far from being all inferior to those in the 'exact' sciences, range from those which are indeed flimsily based to inferences that are now as well established as commonly accepted results in standard physics.

K. E. Bullen, The Earth's density (1975)

7.1 Generalities

All the information we have about the inaccessible interior of the Earth is embodied in Earth models, which if they are well constrained by observations and physical laws, are, at least in some respects, open to as little doubt as accepted tenets of, for instance, astronomy.

The previous chapters were devoted to laying the groundwork of the physics and thermodynamics that apply to the materials constituting the deep Earth, emphasizing the contribution of laboratory experimentation. We are now in a position to summarily present the recent view of the inner Earth that results from the conjunction of these physical constraints with a corpus of ever-improving geophysical observations.

We will follow the traditional, and convenient, habit of separately considering seismological, thermal, and compositional (mineralogical) Earth models. It must, of course, be kept in mind that they strongly interact (Fig. 7.1).

The seismological models are based on velocity–depth profiles determined from the travel-time–distance curves for seismic waves and on periods of free oscillations (see Bullen and Bolt 1985, and, for a clear elementary presentation, Bolt 1982). Due to the development of world-wide networks of three-component broad-band seismographs, there are more and more data, of better and better quality. At the initiative of the International Association of Seismology and Physics of the Earth's Interior, a preliminary reference Earth model (PREM) was recently set up (Dziewonski and Anderson 1981) and at the present time, it is considered to be the best global seismological model available. Seismological models yield density, pressure, and elastic moduli as functions of depth, and before introducing the PREM model, we will deal with the fundamentals of density profiles determination.

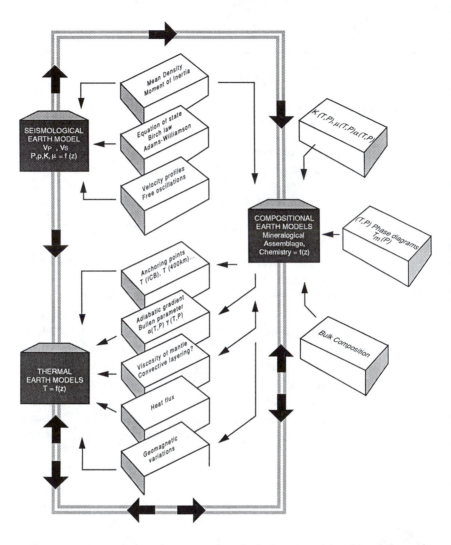

Figure 7.1. Interrelations between seismological, compositional, and thermal Earth models.

Thermal models necessarily depend on experimentally determined thermodynamic parameters, as well as observations of heat flux and geomagnetic variations. The temperature–depth profile (geotherm) has a strong influence on the compositional models since we need to know the temperature at a given pressure to infer from the experimentally determined phase diagrams which minerals are stable.

The compositional models, constrained by the density and velocity profiles from seismological models, in turn act on the thermal models by anchoring the geotherm and by allowing or forbidding convective layering, thus making the introduction of thermal boundary layers necessary or not.

7.2 Seismological models

7.2.1 Density distribution in the Earth

Knowing the mass of the Earth $M = 5.974 \times 10^{24}$ kg and its mean radius $R = 6371$ km, it is obvious that its mean density $\bar{\rho} = 5.515$ is higher than the average density (2.7 to 3.3) of the rocks found at the surface of the Earth. Besides, the moment of inertia of the Earth about its rotation axis, determined from flattening and precession measurements, is $I = 0.33MR^2$, smaller than the value $0.4MR^2$ that would obtain for a homogeneous sphere of constant density, thus pointing to a concentration of mass near the center of the Earth. It could, of course, be entirely due to an increase in density with depth as the rocks are compressed. It is therefore necessary to examine the variation of density with depth, due to compression alone, of an isochemical material; it will turn out that it is insufficient to account for all of the mass concentration toward the center.

We assume that the compression is adiabatic, that is, that there is no exchange of heat, which could cause temperature variations and add a thermal expansion contribution to the density variations with pressure. We also assume that the Earth is in hydrostatic equilibrium and spherically symmetrical, hence

$$dP = -\rho g dr \qquad (7.1)$$

where P is the pressure at radius r or depth z ($r = 6371\,\mathrm{km} - z$) and ρ and g are the density and acceleration of gravity at radius r, respectively, with

$$g = Gmr^{-2} = 4\pi Gr^{-2} \int_0^r \rho r^2 \, dr \qquad (7.2)$$

where m is the mass of the sphere of radius r and density ρ and $G = 6.66 \times 10^{-11}\,\mathrm{Nm^2kg^{-2}}$ is the gravitation constant. Hence

$$\frac{dP}{dr} = -4\pi G\rho r^{-2} \int_0^r \rho r^2 \, dr \qquad (7.3)$$

By definition of the (adiabatic) bulk modulus K and of the seismic paramater Φ, we have:

$$\frac{d\rho}{dP} = \frac{\rho}{K} = \Phi^{-1} \qquad (7.4)$$

Hence, with (7.3),

$$\frac{d\rho}{dr} = -4\pi GK^{-1}\rho^2 r^{-2} \int_0^r \rho r^2 \, dr \tag{7.5}$$

or

$$\frac{d}{dr}\left(r^2 K\rho^{-2}\frac{d\rho}{dr}\right) = -4\pi Gr^2\rho \tag{7.6}$$

N.B.   Remembering that

$$\frac{dU}{dr} = -g = \frac{dP}{dr}\rho^{-1}$$

where U is the gravitational potential, and with the definition of the Laplacian in spherical coordinates

$$\nabla^2 U \equiv \frac{1}{r^2}\frac{d}{dr}\left(r^2\frac{dU}{dr}\right)$$

we see that equation (7.6) is in fact Poisson's equation:

$$\nabla^2 U = 4\pi G\rho$$

Using an equation of state of the form

$$K = C\rho^n \tag{7.7}$$

where C and n are constants, we obtain *Emden's equation* (first established to calculate the pressure inside stars):

▶
$$\frac{d}{dr}\left(r^2\rho^{n-2}\frac{d\rho}{dr}\right) = -A^2 r^2\rho \tag{7.8}$$

with $A^2 = 4\pi G/C$.

We see, from (7.7), that

$$n = \frac{d\ln K}{d\ln\rho} = \frac{dK}{dP} = K_0' \tag{7.9}$$

As we have seen before, $n \approx 4$ for the second-order Birch–Murnaghan equation.

Note that Laplace (1825), assuming that the derivative of the pressure with respect to the density was proportional to the density

$$\frac{dP}{d\rho} = C\rho \tag{7.10}$$

had directly obtained equation (7.8) with $n = 2$, as immediately follows from (7.10). In this case Emden's equation (7.8) has a solution of the form $\rho = \rho_0(Ar)^{-1}\sin(Ar)$, but it has no simple solution in the general nonlinear case.

Still assuming that the interior of the Earth is homogeneous and adiabatic, Williamson and Adams (1923) did not introduce an a priori equation of state such as (7.7) into (7.5) to obtain a differential equation which turns out to be difficult to resolve for $\rho(r)$. They kept equation (7.5) under the form known as *Adams and Williamson's equation:*

$$\blacktriangleright \qquad \frac{d\rho}{dr} = -g\rho^2 K^{-1} = -g\rho\Phi^{-1} \qquad (7.11)$$

or

$$\ln\left(\frac{\rho}{\rho_0}\right) = -\int_{r_0}^{r} g\rho K^{-1}\,dr = -\int_{r_0}^{r} g\Phi^{-1}\,dr \qquad (7.12)$$

where ρ_0 is the density at the surface of the Earth ($r = r_0$).

Equation (7.12) relating ρ and K for a given value of r is indeed an equation of state.

Using *P*- and *S*-wave velocity profiles determined from time–distance curves and starting from the surface with initial density 3 and 3.5, Williamson and Adams obtain a density–depth profile by approximation and repeated graphical integration, layer by layer. They find that the density variation due to compression alone accounts for the density profile in the lower mantle but that the density does not increase fast enough to make the mean density equal to 5.5. They conclude that "It is therefore impossible to explain the high density of the Earth on the basis of compression alone. The dense interior cannot consist of ordinary rocks compressed to a small volume; we must therefore fall back on the only reasonable alternative, namely, the presence of a heavier material, presumably some metals, which, to judge from its abundance in the Earth's crust, in meteorites and in the Sun, is probably iron."

The pressure–depth profile follows immediately by integration of equation (7.1). Williamson and Adams find the pressure at the center of the Earth $P_c = 3.18$ Mbar, which can be compared to the values $P_c = 3.08$ Mbar found by Laplace and $P_c = 3.64$ Mbar of the PREM model. Departure from the conditions of homogeneity and adiabaticity, hence from the conditions of validity of Adams and Williamson's equation can be expressed by the *Bullen parameter* η, defined in the following fashion. The Adams–Williamson equation (7.11) can be written:

$$-\Phi\rho^{-1}g^{-1}\frac{d\rho}{dr} = 1 \qquad (7.13)$$

If the conditions of adiabaticity or homogeneity are not fulfilled, the Bullen parameter is defined as being the value of the left-hand side member of

(7.13), no longer equal to 1, since ρ then corresponds to a nonhomogeneous or nonadiabatic region:

$$-\Phi\rho^{-1}g^{-1}\frac{d\rho}{dr} = \eta \tag{7.14}$$

The pressure derivative of the bulk modulus can be expressed in terms of the Bullen parameter; starting with the definition of the seismic parameter $K = \rho\Phi$ and taking the derivative of K with respect to pressure, we obtain, with (7.14),

$$\frac{dK}{dP} = \left(\Phi\frac{d\rho}{dr} + \rho\frac{d\Phi}{dr}\right)\frac{dr}{dP} = \eta - \frac{1}{g}\frac{d\Phi}{dr} \tag{7.15}$$

which yields another expression for Bullen's parameter:

$$\blacktriangleright \qquad\qquad \eta = \frac{dK}{dP} + \frac{1}{g}\frac{d\Phi}{dr} \tag{7.16}$$

Another interesting expression for η immediately follows from (7.14), if we write, using the model of (7.13),

$$\rho^{-1}g^{-1}\frac{d\rho}{dr} = -\Phi_{obs}^{-1} \tag{7.17}$$

where Φ_{obs} is the actual, observed, value of the seismic parameter, and ρ is the density of the nonhomogeneous or nonadiabatic region; we then have

$$\eta = \frac{\Phi}{\Phi_{obs}} = \frac{K}{K_{obs}} \tag{7.18}$$

We then see that there is a straightforward physical meaning to Bullen's parameter:

If $\eta = 1$, as in most of the lower mantle, the material can be thought to be homogeneous and the temperature gradient is equal to the adiabatic gradient.

If $\eta > 1$, the material is apparently more compressible than in the homogeneous adiabatic conditions; this can happen when there are phase transitions, as in the mantle transition zone, where $1.7 < \eta < 2$.

If $\eta < 1$, the material is apparently less compressible than in the homogeneous adiabatic conditions and this can be due to the effect of thermal expansion resulting from nonadiabaticity. This is the case in the thermal boundary layers, for example, at the top of the mantle, where $\eta = 0.7$.

In the nonadiabatic case, the Adams–Williamson equation can be corrected by the thermal expansion term, due to the superadiabatic gradient:

$$\frac{d\rho}{dr} = -g\Phi^{-1}\rho + \alpha\rho\left(\frac{dT}{dr} - \frac{g\alpha T}{C_p}\right) \tag{7.19}$$

7.2.2 The PREM model

Seismological Earth models typically use the velocity–depth profiles and an equation of state relating ρ to K (or Φ or v_P) to obtain density, pressure, and elastic moduli profiles. We will deal here only with the most recent and complete model, the preliminary reference Earth model PREM (Dziewonski and Anderson 1981).

The Earth is divided into radially symmetrical shells separated by convenient seismological discontinuities, of which the principal are situated at depths of 400, 670, 2890, and 5150 km, corresponding to the seismic boundaries between uppermost mantle and transition zone, upper and lower mantle, mantle and core, and outer and inner core, respectively.

It is assumed that Adams and Williamson's equation is justified in each region from the center up to the 670 km discontinuity and that Birch's law $\rho = a + bv_P$ can be applied in the upper mantle. The Earth's mass and its moment of inertia are given.

Starting values are assigned to the density below the crust ($\rho = 3.32$), at the base of the mantle ($\rho = 5.5$), and to the density jump between inner and outer core ($\Delta\rho = -0.5$).

The density at the center of the Earth and the jump in density at 670 km are calculated and found to be $\rho_c = 12.97$ and $\Delta\rho = -0.35$, respectively. The starting density distribution is then known.

The observed values entering the model are the travel times of P and S body waves with a period of 1s and the periods of free oscillations, together with the attenuation factors. The starting model is defined by a set of five functions of radius: the velocities v_P and v_S, the density, and the attenuation factors in shear and compression. The inverse problem is solved simultaneously for elastic and anelastic parameters, and perturbations are introduced into the starting model to satisfy the data. Elastic anisotropy is introduced in the uppermost 200 km of the upper mantle. The parameters of the final model are given as polynomials in r, or tabulated (see the appendix).

The velocity, density, and pressure profiles of the PREM model are given in Figure 7.2 and Figure 7.3; the variations of the seismic parameter Φ and of Poisson's ratio with depth are given in Figure 7.4 and Figure 7.5.

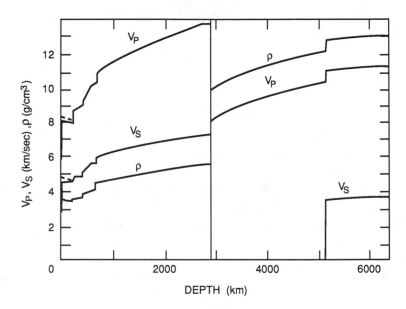

Figure 7.2. PREM model: Seismic velocities and density profile (after Dziewonski and Anderson 1981).

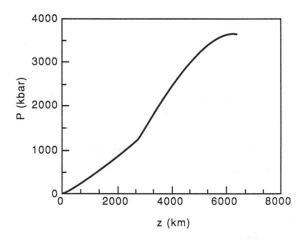

Figure 7.3. PREM model: Pressure profile.

Note that the Poisson's ratio of the outer core is equal to 0.5 as expected for a liquid, but the Poisson's ratio of the inner core is also quite high ($\nu = 0.44$). Various explanations have been given, including the possibility of a noncrystalline inner core. However, such conclusions are unnecessary,

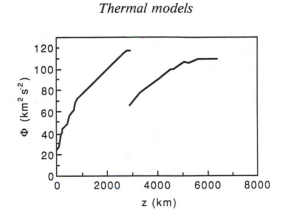

Figure 7.4. PREM model: Seismic parameter profile.

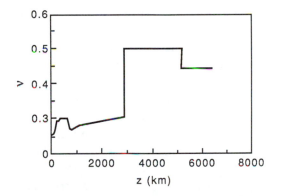

Figure 7.5. PREM model: Poisson's ratio profile.

since a high Poisson's ratio is not unusual in metals, for example, 0.42 for gold and 0.45 for indium. Falzone and Stacey (1980) give an explanation of the high Poisson's ratio at high pressure in terms of the second-order theory of elasticity.

7.3 Thermal models

7.3.1 Sources of heat

For a discussion of the sources of heat and temperature in the Earth, the reader will profitably refer to Verhoogen's (1980) illuminating little book *Energetics of the Earth*.

The heat flux coming from inside the Earth can be measured at the surface; its mean value is about 80 mW/m^2 or 4×10^{13} W for the whole Earth. This is, of course, a boundary condition for any thermal model, but to infer the temperature profile, one must also have some idea of the sources of the heat that is transported through the mantle and finally radiated out at the surface.

How much of the heat is original and how much is currently produced in the Earth? In other words, is the Earth still cooling from an original hot state, as was widely thought in the last century, or are there active sources of heat inside? We now know that radioactivity is the major heat source, but others may exist, and whether the Earth is still cooling is a matter of current debate. Let us briefly review the possible sources of heat.

i. The original heat. It is the heat content of the Earth in the early stages of its history. It is essentially accretional heat, due to the dissipation of the gravitational energy when planetesimals bombarded the surface of the growing Earth, which eventually partly melted. There also was a contribution of short-lived, now extinct, radioactive elements, such as Al^{26}. During the differentiation stage that ensued, gravitational energy was again released when droplets of liquid iron or iron–sulfur eutectic trickled down to form the core. It is generally thought that the core was formed in a relatively short time, ending about 0.5 billion years after the formation of the Earth; the corresponding heat therefore can be said to be "original." It is believed, although not universally, that the original heat contributes little to the thermal budget, with the exception of the heat stored in the liquid core.

ii. Radiogenic heat. Decay of the radioactive elements present in the mantle is the main source of heat in the Earth. The principal radioactive elements are U^{235}, U^{238}, and Th^{232}, whose decay eventually gives Pb^{207}, Pb^{206}, and Pb^{208}, respectively, and K^{40}, whose decay gives Ca^{40} and A^{40}. The heat production per mass unit of each element is well known but their concentration in the Earth is much less certain. One estimate (Verhoogen 1980) leads to the approximate lower bound of 2.4×10^{13} W for the total radiogenic production of heat in the mantle, that is, about 60 percent at least of the total output of heat.

iii. Other sources of heat. They include tidal dissipation in the solid Earth, which is negligible at the present time, although it may have been important in the past, when the Moon was closer to the Earth; frictional dissipation in the convecting mantle; and latent heat released in exothermal phase transitions (e.g., olivine–spinel). All these contributions are unimportant when compared to that of radiogenic heat. However, a non-negligible contribution to the heat output of the core is made by the latent heat released during crystallization of the inner core and by the gravitational energy released as particles of solid iron fall toward the inner core and as the mantle falls in on the shrinking core.

N.B. The latent heat released during crystallization of the inner core is $L = T_f \Delta S_f$. With the values calculated by Poirier (1986) for the temperature of the inner core boundary ($T_f = 5000$ K) and the melting entropy ($\Delta S_f = 5.83$ J mol^{-1}K^{-1}) and taking a molar volume at the pressure of the inner core boundary of 4.38 cm^{3}mol^{-1} and density of 12, we find $L = 5.55 \times 10^5$ J Kg^{-1}.

The heat flux from the core into the lower mantle is another boundary condition of the convective problem. It is the sum of the term representing the cooling of the whole core $Q_c \approx 2.6 \times 10^{12}$ W, the term due to the crystallization of the inner core $Q_c \approx 0.34 \times 10^{12}$ W and the gravitational energy term $Q_c \approx 0.66 \times 10^{12}$ W (Verhoogen 1980). The total heat output of the core (input into the lower mantle) is therefore $Q_c \approx 3.6 \times 10^{12}$ W, about 10 percent of the total heat output of the Earth.

7.3.2 Heat transfer by convection

We have seen (Sec. 6.6) that, due to the opacity of the iron-bearing minerals under high pressure, radiative transfer of heat is most probably negligible in the Earth's mantle. The two remaining mechanisms for heat transfer are then conduction, consisting in heat transport by thermal vibrations of the mineral lattices, and convection, in which the heat-containing matter is bodily transported in a fluidlike manner: On the time scale of the geophysical phenomena (e.g., plate tectonics) the mantle can be considered as a fluid endowed with a very high viscosity (10^{21}–10^{22} Pa s).

It is completely beyond the scope of this book to deal with the fluid mechanics of convection. However, we will give the physical bases of the convective phenomenon in the simpler case of a fluid heated from below in a gravitational field (Rayleigh–Bénard convection) (see Tritton 1977; Turcotte and Schubert 1982).

Let us consider a laterally infinite layer of fluid of density ρ in the gravitational field of the Earth, bounded by plane surfaces at $r = 0$ and $r = h$ (Fig. 7.6a). The upper surface is maintained at a fixed temperature T_0, while the lower surface is maintained at $T_1 > T_0$, thus establishing a temperature gradient $\nabla T = (T_1 - T_0)/h$ through the fluid. Let us now consider

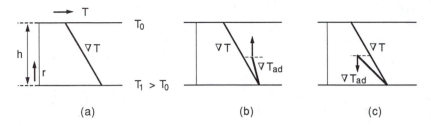

Figure 7.6. Convective instability in a sheet of fluid heated from below. (a) Temperature profile. (b) The temperature profile is superadiabatic: a parcel of fluid displaced upward is lighter than the surrounding fluid and keeps going up (convective instability). (c) The temperature profile is subadiabatic: a parcel of fluid displaced upward is heavier than the surrounding fluid and falls back (stable layering).

a small parcel of fluid at $r = 0$ and $T = T_1$ and let it rise rapidly by δr. It undergoes an adiabatic decompression and its temperature is lowered by

$$\delta T = -\nabla T_{ad}\delta r \qquad (7.20)$$

where ∇T_{ad} is the *adiabatic gradient,* given by (2.57). Since $dP = \rho g dz = -\rho g dr$, we have

$$\nabla T_{ad} = \frac{dT}{dz} = \frac{g\alpha T}{C_p} \qquad (7.21)$$

Note that we have

$$\nabla T_{ad} = g\gamma_{th}\Phi^{-1}T \qquad (7.22)$$

where γ_{th} is the thermodynamic Grüneisen parameter and Φ is the seismic parameter; the adiabatic gradient in the lower mantle is about 0.3 K/km.

If the temperature gradient in the fluid is subadiabatic, that is, $|\nabla T| < |\nabla T_{ad}|$ (Fig. 7.6b), the parcel of fluid is cooler, hence denser, than the surrounding fluid and sinks again. The fluid is stratified and stable with respect to convection. Heat is transported by conduction and the heat flux (per unit time and unit area) is

$$H_c = -k\nabla T = k\frac{T_1 - T_0}{h} \qquad (7.23)$$

where k is the thermal conductivity of the fluid.

If the temperature gradient in the fluid is superadiabatic, that is, $|\nabla T| > |\nabla T_{ad}|$ (Fig. 7.6c), the parcel of fluid is warmer and lighter than the surrounding fluid; it is buoyant and will go on rising. The situation is unstable. The criterion for the onset of convection is therefore that the temperature gradient be superadiabatic, or that the Bullen parameter be smaller than 1. The density of a fluid parcel displaced by δr upward is $\rho + \delta r (d\rho/dr)_{ad}$, whereas the density of the surrounding fluid is $\rho + \delta r (d\rho/dr)$. The gravitational force on the parcel is therefore $-g\delta r[(d\rho/dr)_{ad} - d\rho/dr]$ and the equation of motion of the parcel is

$$\rho \frac{d^2 \delta r}{dt^2} + g\delta r \left[\left(\frac{d\rho}{dr} \right)_{ad} - \frac{d\rho}{dr} \right] = 0 \qquad (7.24)$$

The fluid parcel oscillates about its original position with the *Brunt–Väisälä frequency N* (see Tritton 1977),

$$N = \left\{ \frac{g}{\rho} \left[\left(\frac{d\rho}{dr} \right)_{ad} - \frac{d\rho}{dr} \right] \right\}^{1/2} \qquad (7.25)$$

The Brunt–Väisälä frequency is related to the Bullen parameter η and is sometimes used in the geophysical literature:

$$N^2 = \frac{g^2}{\Phi} (\eta - 1) \qquad (7.26)$$

We see that, if the gradient is superadiabatic ($\eta < 1$), the frequency is imaginary and a convective instability starts and can grow exponentially if the physical properties of the fluid (essentially its viscosity) are such as to allow convection. The condition $\eta < 1$ is necessary but not sufficient for convection. A tighter criterion is given by the dimensionless *Rayleigh number,* defined, in the case of a fluid heated from below, by

$$R_a = \frac{g\alpha h^3 \Delta T}{\nu \kappa} \qquad (7.27)$$

where $\nu = \eta/\rho$ is the kinematic viscosity of the fluid, κ is its thermal diffusivity, α is its thermal expansion coefficient, and $\Delta T = T_1 - T_0$.

The Rayleigh number measures the relative importance of the buoyancy force $g\alpha\Delta T$, favoring convection, and the viscosity drag force $\nu\nabla^2 V$, hindering convection (see 7.18).

$$R_a = \frac{g\alpha\Delta T}{\nu\nabla^2 V} \qquad (7.28)$$

If we scale velocity V to the value U, length to L, and time to L/U, we have $\nabla^2 V \approx UL^{-2}$ and $\kappa \approx L^2(L/U)^{-1} = UL$, hence $U \approx \kappa L^{-1}$ and $\nabla^2 V \approx \kappa L^{-3}$. (7.28) is therefore equivalent to (7.27).

In the case of a fluid heated from within, the Rayleigh number is (see Turcotte and Schubert 1982)

▶
$$R_a = \frac{g\alpha\rho q h^5}{\nu\kappa k} \tag{7.29}$$

where k is the thermal conductivity of the fluid and q is the rate of internal heat production.

A linear stability analysis shows that perturbations can grow exponentially when the Rayleigh number is greater than the critical Rayleigh number $R_a^{(c)}$. For fluids heated from below as well as from within, $R_a^{(c)} \approx 2000$.

N.B.   Let us calculate an order of magnitude for the Rayleigh number of the internally heated Earth's whole mantle, using as values of the various parameters:

$$\rho \approx 4\times10^3 \text{ kg m}^{-3}, \quad \alpha \approx 3\times10^{-5}\text{K}^{-1}, \quad \kappa \approx 0.01 \text{ cm}^2\text{s}^{-1}, \quad \eta \approx 10^{21} \text{ Pa s},$$

$$g \approx 10 \text{ ms}^{-2}, \quad \nu = \eta/\rho \approx 3\times10^{17} \text{ m}^2\text{s}^{-1}, \quad h \approx 3000 \text{ km},$$

$$k \approx 4 \text{ Wm}^{-1}\text{K}^{-1}, \quad q \approx 9\times10^{-12} \text{ Wkg}^{-1}.$$

We find a Rayleigh number of about 2×10^9, considerably above the critical value.

The *Nusselt number Nu* measures the thermal efficiency of heat transfer. It is equal to the ratio of the total heat flux H to the heat flux that would be transported by conduction under the same conditions. With (7.23), we have:

$$Nu = \frac{Hh}{k\Delta T} \tag{7.30}$$

For $R_a > R_a^{(c)}$, we have $Nu > 1$.

Under steady state conditions, convection in a viscous fluid is controlled by the coupled differential equations for conservation of momentum and for transport of heat:

$$\rho\frac{D\mathbf{v}}{Dt} = \Sigma \mathbf{F} \tag{7.31}$$

$$\frac{DT}{Dt} = \kappa\nabla^2 T \tag{7.32}$$

N.B.   The symbol D/Dt represents the material (or convective) derivative, which must be used when considering the variation with time of a quantity y in a fluid moving with a velocity $\mathbf{v}$.

$$\frac{Dy}{Dt} = \frac{\partial y}{\partial t} + \Sigma_i \frac{\partial y}{\partial x_i}\frac{\partial x_i}{\partial t} = \frac{\partial y}{\partial t} + \Sigma_i \frac{\partial y}{\partial x_i}v_i$$

x_i and v_i ($i = 1, 2, 3$) are the components of the position and velocity vectors, respectively. We can write:

$$\frac{Dy}{Dt} = \frac{\partial y}{\partial t} + \mathbf{v} \cdot \nabla y$$

The term $\mathbf{v} \cdot \nabla y$ is called the advective term.

In the equation of conservation of momentum for a viscous fluid, the forces $\mathbf{F}$ to be taken into account are the force resulting from a pressure gradient and the force needed to overcome viscosity (see equation (6.18)). We can write (7.31) under the form known as the *Navier–Stokes equation:*

$$\frac{\partial \mathbf{v}}{\partial t} + (\mathbf{v} \cdot \nabla)\mathbf{v} = -\frac{1}{\rho} \nabla P + \mathbf{g} + \nu \nabla^2 \mathbf{v} \qquad (7.33)$$

The heat transport equation is written as

$$\frac{\partial T}{\partial t} + \mathbf{v} \cdot \nabla T = \kappa \nabla^2 T \qquad (7.34)$$

In the case of vigorous convection, the term corresponding to the diffusion of heat, $\kappa \nabla^2 T$, is negligible compared to the advective term, $\mathbf{v} \cdot \nabla T$. The heat is transported with matter and has no time to diffuse far; the convection is therefore adiabatic.

At the top and bottom of the convection cells, where heat is transferred into the convecting fluid or out of it, or between thermally coupled but isolated convecting systems, there must exist a *thermal boundary layer* through which heat is transferred by conduction and where the gradient is highly superadiabatic.

7.3.3 Convection patterns in the mantle

Continental drift and plate tectonics are proof enough that the upper mantle convects. There is no direct evidence for convection in the lower mantle and it was thought for some time that the viscosity increase with pressure could be so high as to prevent convection (MacKenzie 1967). This view, which partly resulted from not taking into account the decrease of activation volume for creep with increasing pressure, is no longer currently entertained: The Rayleigh number for the lower mantle is supercritical and the Bullen parameter is close to 1 in most estimates, pointing to a convective mantle. The controversy now centers, not on whether the lower mantle convects or not, but on whether it convects together with the upper mantle (whole mantle convection) or separately, with a thermal boundary layer at 670 km (two-layer convection).

The arguments presented by each school of thought for their favorite model and against the other one are generally cogent and persuasive, which makes the question difficult to decide for an unprejudiced observer, all the more so since their proponents usually rely, perforce, on various unverifiable assumptions and on the feeling that the opposite view would lead to improbable values of parameters which are, anyway, very little constrained by observational data. Note also that many of the arguments brought forward merely show that a given observation or calculation is consistent with one style of convection but generally do not prove that another style is excluded.

Reviewing the extant literature on convection, including the experimental and numerical models, could be the subject of a whole book. It will suffice here to briefly present the pro and con arguments for the main contending models and leave the reader to make up her or his own mind. Note that some authors use a temperature profile to defend a convection style, whereas others posit a convection style to derive a temperature profile. Indeed, the two problems are so intimately linked that it is only for the sake of convenience that we deal with them in different paragraphs.

i. Two-layer convection. The principal arguments proposed are as follows:

The deepest earthquakes stop at 670 km and their focal mechanisms imply down-dip compression. It is therefore concluded that the subducting plates cannot penetrate the lower mantle.

The 670-km discontinuity, in addition to being due to a phase transition, is also a compositional and/or chemical boundary (Liu 1979). The density and seismic velocities agree with a lower mantle richer in silica (Anderson and Bass 1986; Duffy and Anderson 1989) and/or richer in iron (D. L. Anderson 1989) than the upper mantle (see Sec. 7.4).

The 670-km discontinuity is a sharp seismic reflector that cannot be produced by a phase transition occurring over a wide depth interval. Hence it must also be a chemical boundary (Lees, Bukowinski, and Jeanloz 1983).

The mantle has the same composition throughout but the convective systems of the upper and lower mantle are isolated by the subducted lithosphere trapped between 600 and 700 km (Ringwood and Irifune 1988). A chemically heterogeneous "mesosphere boundary layer" between the upper and lower mantle could be the source of the Ocean Island Basalts (Allègre and Turcotte 1987).

The upper and lower mantle are two different geochemical reservoirs that separately fractionated to produce the continental crust and the core, respectively (Allègre 1982).

A thermal boundary layer inside the mantle is required in the thermal
model, in order to avoid using an unreasonably high value of the
heat flux from the core into the mantle (Jeanloz and Richter 1979).

ii. Whole-mantle convection. Most of the arguments in favor of this
model are negative, that is, against the two-layer model: They contend
either that it is not required by the data (then, why not accept the simpler
solution?) or that it would lead to unreasonable assumptions.

A compositional difference between upper and lower mantle is not required
(although not excluded) by the seismological data (Weidner 1986).

The analysis of travel times of deep earthquakes shows that there are fast
(colder) regions below 670 km in the prolongation of some subducting slabs which are therefore thought to penetrate into the lower
mantle (Fischer, Jordan, and Creager 1988).

The seismic constraints on the sharpness of the 670-km discontinuity are
weak (Muirhead 1985).

The post-spinel phase transition at 670 km is sharp and could account for
a good seismic reflectivity; a compositional difference between upper
and lower mantle is not required (Ito and Takahashi 1989).

The existence of a thermal boundary layer at 670 km, where heat is transferred by conduction between the isolated convection cells, implies
a high temperature gradient, hence a temperature in the lower mantle higher by about 800°C than in the case of whole-mantle convection. As the lower mantle viscosity is thermally activated, a higher
temperature implies a viscosity of about 6×10^{16} Pa for the lower
mantle (Kenyon and Turcotte 1983), whereas the viscosity deduced
from post-glacial rebound is of the order of 10^{21} Pa (Peltier and
Jarvis 1982) (Fig. 7.7). Note, however, that in the absence of experimental data on the high-temperature creep of the lower-mantle material, any kind of extrapolated or otherwise calculated viscosity of
the lower mantle is largely a matter of personal choice.

Three-dimensional spherical models of whole-mantle convection (Bercovici, Schubert, and Glatzmaier 1989) predict upwelling plumes
and downwelling sheets closely resembling the circum-Pacific sub-
duction ring (but so would presumably 3-D models of two-layer
mantle convection). Indeed, seismic tomography of the lower mantle shows a ring of high-velocity material, which projects to the surface as a circum-Pacific ring and might be the continuation of the
subducting slabs (Olson, Schubert, and Anderson 1990; Olson, Silver, and Carlson 1990).

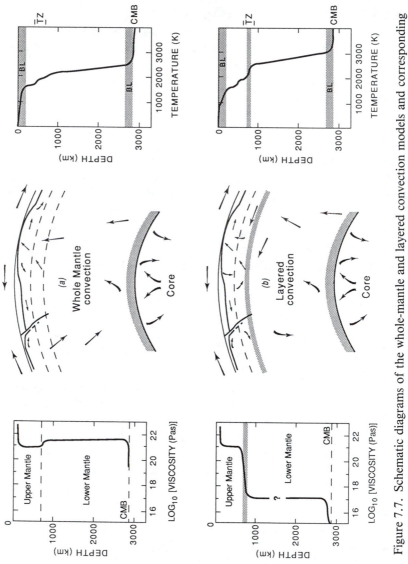

Figure 7.7. Schematic diagrams of the whole-mantle and layered convection models and corresponding temperature and viscosity profiles (after Peltier and Jarvis 1982).

Interesting observations of strong correlations among the geoid, the surface topography, and the seismic velocity anomalies in the lower mantle have recently been presented (Hager et al. 1985; Cazenave, Souriau, and Dominh 1989). The geoid anomalies observed at the surface result from the interior density contrasts (hot and cold matter) driving the convective flow and from the deformation of the boundaries caused by the flow; the magnitude and sign of the total effect depend on the viscosity profile and the style of convection. However, although the results of the analyses are compatible with one style of convection, they are not incompatible with the other.

Christensen (1984) modeled a variety of convecting styles and showed that "with the present uncertainties about the 670 km discontinuity a variety of convection styles are possible." He noted in particular that a pure phase transition boundary could produce a kind of leaky double-layer convection if the Clapeyron slope was as low as -6 MPa/K, which he thought at the time to be unlikely but not entirely unreasonable; it is in fact not significantly different from the most recent experimental value of -3 MPa/K (Ito and Takahashi 1989).

7.3.4 Geotherms

Temperature profiles (geotherms) are usually anchored in the depths of seismic discontinuities identified with phase transitions whose P, T boundaries are experimentally known or extrapolated. The principal discontinuities used are the inner core boundary, identified with the freezing of the liquid core iron alloy, and the 670-km discontinuity, identified with the post-spinel transition. Starting from these anchoring points, the geotherms follow an adiabat in the homogeneous regions where the Bullen parameter is close to 1 (lower mantle and outer core), the adiabatic gradient often being determined using a value of the acoustic Grüneisen parameter compatible with a seismological Earth model. The uncertainty is seldom claimed to be less than a few hundred degrees and often is of the order of 1000 K. The temperatures at various characteristic depths, determined in recent studies, are given in Table 7.1 and the geotherms are shown in Figure 7.8.

In the lithosphere, the geotherm is often derived from pyroxene thermobarometry measurements made on peridotite xenoliths (e.g., Mercier 1980).

In the transition zone, the most recent estimates using the phase diagram of the system $Mg_2SiO_4-Fe_2SiO_4$ (Ito and Katsura 1989) give a temperature of 1400°C at 350 km and of 1600°C at 655 km.

Table 7.1. *Temperatures inside the Earth (in Kelvins)*

z (km)	W72	V80	A82	SS84	BS85	BM86	P86	W87	IK89
100		1273	1450						
150		1473							
350									1400
371			1662						
655									1600
671−			1830	1970	1873				
671+			1980	2300					
1300	2800								
2571	3300		2814						
2886−			2937	3000	2773				
2886+			3637	3800	3573	3800	3800		
5156			4676	5000		5000	5000	6600	
6371			4805			5000		6600	

Source: W72: Wang (1972b). V80: Verhoogen (1980). A82: O. L. Anderson (1982). SS84: Spiliopoulos and Stacey (1984). BS85: Brown and Shankland (1981). BM: Brown and Mc-Queen (1986). P86: Poirier (1986). W87: Williams, Jeanloz, et al. (1987). IK89: Ito and Katsura (1989).

In the lower mantle, Shankland and Brown (1985), starting from the temperature at 670 km (1600°C) and using a seismological adiabatic gradient, arrive at a temperature of 3300°C on the outer core side of the core–mantle boundary, in agreement with the temperature derived from their measurements of shock melting in iron, provided one assumes that the D'' zone at the base of the mantle is a thermal boundary layer with $\Delta T = 800$ K. Under these conditions, they do not find it necessary to introduce a thermal boundary layer at 670 km.

Spiliopoulos and Stacey (1984) start upward from the temperature at the inner core boundary, using $\Delta T = 800$ K in the D'' zone, and meet at 670 km a profile extrapolated down from the Mg_2SiO_4 phase transition. The temperature difference of 300 K they find is thought to be too small for a thermal boundary layer to be stable. They conclude that there is probably no boundary layer at 670 km, although it cannot be ruled out.

A temperature difference of 800 K across the D'' zone corresponds to a gradient of 8 K/km over 100 km; this usually adopted value is derived from Fourier's law assuming a heat flux from the core of 0.032 Wm^{-2} and a thermal conductivity of the lower mantle of $k = 4$ $Wm^{-1}K^{-1}$. Brown (1986), taking into account the variation of k with temperature

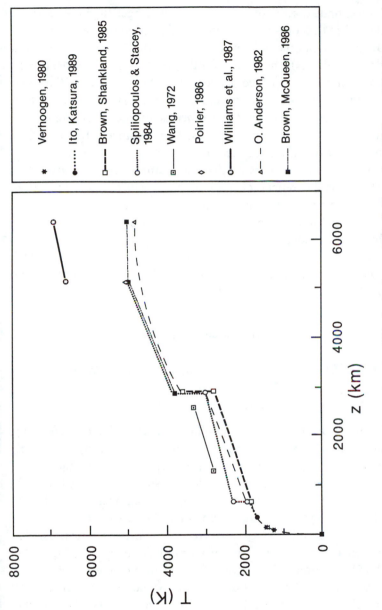

Figure 7.8. Geotherms (Wang 1972b; Verhoogen 1980; O. Anderson 1982; Spiliopoulos and Stacey 1984; Brown and Shankland 1981; Brown and McQueen 1986; Poirier 1986; Williams, Jeanloz, et al. 1987; Ito and Katsura 1989). For all these temperature profiles, the error bar is about ±800 K.

Verhoogen, 1980

Ito, Katsura, 1989

Brown, Shankland, 1985

Spiliopoulos & Stacey, 1984

Wang, 1972

Poirier, 1986

Williams et al, 1987

O. Anderson, 1982

Brown, McQueen, 1986

and pressure, finds that it could be three times larger than the currently adopted value at the core–mantle boundary, and that the need for a thermal boundary layer at the base of the mantle then disappears, unless one assumes an unrealistically high value for the heat flux from the core (higher than the heat flux at the surface of the Earth). Note, however, that there is a trade-off between the thermal boundary layers at 670 km and at the core–mantle boundary and that the geotherms, estimated assuming a homogeneous mantle, cannot really be used as arguments for or against whole–mantle or two-layer convection.

Despite the large uncertainty on the temperatures, most recent models, whether based on theoretical calculations (Poirier 1986) or extrapolations of the melting temperature of iron at the inner core boundary (O. L. Anderson 1982; Spiliopoulos and Stacey 1984; Brown and McQueen 1986) generally agree on a temperature at the inner core boundary close to 5000 K. The consensus is, however, not unanimous since Williams et al. (1987) extrapolate the melting point of iron to a much higher temperature and propose a temperature of 6600 K at the inner core boundary.

7.4 Mineralogical models

7.4.1 Phase transitions of the mantle minerals

i. Generalities on phase transitions

Nature's schemes for building up crystalline edifices from atoms are many, and compounds with a given crystal structure are generally stable only in a limited region of the temperature–pressure plane, where their Gibbs free energy is minimal. Even a simple element such as iron has four solid phases and the number of structures a polyatomic mineral can adopt at various pressures and temperatures can indeed be very large, especially if one considers the additional degree of freedom afforded by the possibility of partial replacement of one kind of atom by another (e.g., Mg by Fe, Si by Al). Descriptions of the atomic architecture of the mineral structures can be found in the specialized volumes of the "Reviews in Mineralogy" or in such books as Wells (1985) or Muller and Roy (1974) for ternary oxides.

The number of thermodynamic degrees of freedom of a system (or *variance*) is the number of intensive parameters capable of independent variation. It is given by the Gibbs phase rule (see Callen 1985):

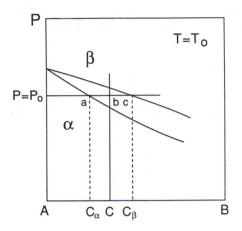

Figure 7.9. Lever rule: At $T = T_0$ and $P = P_0$ and for a composition C, for example, x mol% of component B, the mole fraction x_β of high pressure phase β, of composition C_β, and the mole fraction x_α of low pressure phase α, of composition C_α, are in the proportion $x_\alpha/x_\beta = bc/ab$.

$$\mathcal{V} = c + 2 - \phi \tag{7.35}$$

where $\mathcal{V}$ is the variance, c is the number of independent components, and ϕ is the number of phases in equilibrium.

In the simplest case of one-component systems, when the only intensive variables are temperature and pressure, the equilibrium between two phases is univariant and the domains of stability of two phases in the P–T plane are separated by a line boundary; that is, the equilibrium pressure and temperature cannot be varied independently. The map of the stability domains of the phases in the P–T plane is the *phase diagram*. For binary (two-component) systems, one usually considers sections of the three-dimensional phase diagram by planes of constant T or P. The phase-diagram boundaries are replaced by two-phase loops and at a given P (or T), for a given global composition, there are equilibrium mixtures of two phases. The proportion of the phases varies with T (or P) and is given by the "lever rule" (see Callen 1985) (Fig. 7.9). The case of the ternary systems is obviously more complicated and one often uses "pseudo-binary diagrams" (e.g., P, Fe % at constant T, between the compositions $MgSiO_3$ and $FeSiO_3$ in the ternary system MgO–FeO–SiO_2). One must of course beware not to reason on pseudobinary diagrams as if they were binary diagrams.

Phase transitions can be usefully classified according to their *order* in Ehrenfest's sense (see Rao and Rao 1978). For first-order transitions, there is a discontinuity in the first derivatives of the Gibbs free energy, entropy S and specific volume V, hence first-order transitions are accompanied by a volume change ΔV and latent heat $L = T\Delta S$ is absorbed (endothermal transition) or evolved (exothermal transition).

The slope of the phase boundary is given by the Clausius–Clapeyron rule (5.2)

$$\frac{dT_t}{dP} = \frac{\Delta V}{\Delta S} \tag{7.36}$$

already mentioned in the case of melting (melting is a first-order phase transition). Here T_t is the temperature of transition at pressure P.

The specific volume of the high-pressure phase is always smaller than that of the low-pressure phase, and the high-temperature phase always has a higher entropy than the low-temperature phase. In many cases, the high-pressure phase has a lower entropy than the low-pressure phase and its stability field is wider at low temperatures, that is, the slope of the phase boundary is positive. However, at very high pressures, when the increase of the coordinance of the small cations (e.g., silicon going from 4 to 6 coordinance) is accompanied by an increase in the length of certain bonds, the transition toward the high-pressure phase may be accompanied by an increase in entropy. The slope of the phase boundary is then negative. This was predicted by Navrotsky (1980) and verified in the case of the very high-pressure phases of the mantle silicates.

In a first-order phase transition, the phases are physically separated by a surface of discontinuity: the phase boundary. In many cases, the transformation operates by nucleation and growth of one phase at the expense of the other, the growth being effected by diffusion-controlled displacement of the phase boundary.

Martensitic transformations are rapid, diffusionless transformations characterized by crystallographic orientation (topotactic) relations between parent and daughter phase and corresponding to a shear of the lattice.

For second-order phase transitions, the discontinuity in the derivatives of the Gibbs free energy affects only the second derivatives (e.g., specific heat, incompressibility) and there is no coexistence of phases on each side of a phase boundary. These transitions are often displacive transitions, the change in crystal structure corresponding to a mere distortion of the bonds, whereas in reconstructive first-order transitions atoms have to change places.

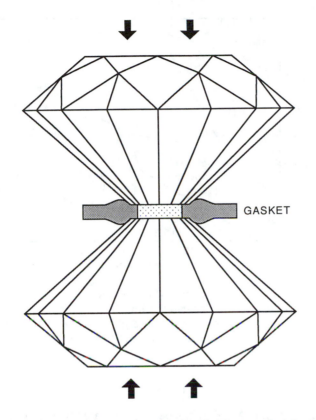

Figure 7.10. Principle of the diamond–anvil cell. The sample (stippled) is contained in a small hole (200 to 400 μm) drilled in a metal gasket, compressed by the diamonds. It can be heated by focusing a laser beam on it.

For more information on the mechanisms and kinetics of the phase transitions in minerals, the reader is referred to Putnis and McConnell (1980). The phase diagrams of elements, oxides, and silicates have been recently reviewed by Liu and Bassett (1986).

In the following sections, we will summarize the state of the art concerning the experimental data on the principal isochemical phase transitions thought to occur in the Earth's mantle, particularly focusing on the polymorphic transitions of $MgSiO_3$ and Mg_2SiO_4. Most of the progress in recent years (Akimoto 1987) has been achieved using two high-pressure techniques: the multi-anvil apparatus, up to about 250 kbar, and the laser-heated diamond–anvil cell (Fig. 7.10), up to 1 Mbar (Ming and Bassett 1974; Bassett 1977; Hemley et al. 1987).

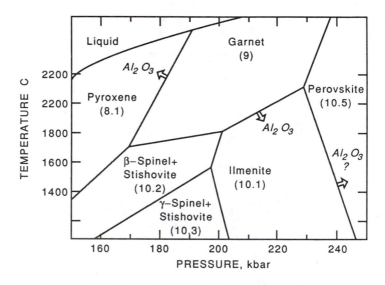

Figure 7.11. *P, T* Phase diagram for MgSiO₃ composition. The approximate velocity of *P* waves (in km/s) is indicated below the names of the mineral phases.
The arrows show the direction in which the phase boundaries are expected to
move when Al₂O₃ is added (after D. L. Anderson 1987b).

ii. The phase transitions of MgSiO₃

The phase diagram of $MgSiO_3$ has been investigated by Liu (1976), Ito and
Yamada (1982), Ito and Navrotsky (1985), Kato and Kumazawa (1985),
Akaogi et al. (1987), Irifune (1987), Sawamoto (1987), and Ito and Taka-
hashi (1989), among others. The resulting synthetic phase diagram in the
P–T plane (D. Anderson 1987b) and the pseudobinary diagram at 1000°C
(Jeanloz and Thompson 1983) are given in Figures 7.11 and 7.12.

The low-pressure phase with composition $MgSiO_3$ is *enstatite,* an orthorhombic pyroxene, with two formula units per unit cell ($Mg_2Si_2O_6$);
its structure can be described as consisting of chains of corner-sharing
SiO_4 tetrahedra, with Mg^{2+} cations in the appropriate sites between the
chains. Enstatite forms a continuous solid solution with the pyroxene
ferrosilite $FeSiO_3$. The upper-mantle material, found in peridotite xenoliths or in ophiolites, has a composition $Mg_{(1-x)}Fe_xSi_2O_6$, with $x \cong 0.1$
in most cases. At higher pressures, the orthorhombic enstatite changes to
monoclinic clinoenstatite. At pressures between about 170 and 190 kbar
and above about 2000 K, the pyroxene transforms into a phase with garnet structure (isolated tetrahedra) and four formula units per unit cell

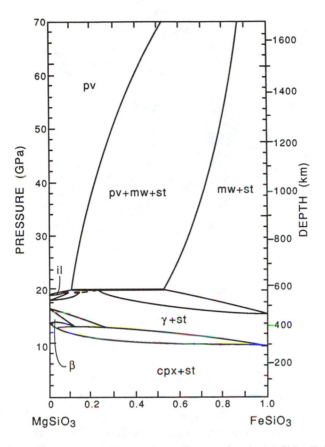

Figure 7.12. Isothermal ($T = 1000°C$) phase diagram for the $MgSiO_3$–$FeSiO_3$ system. The crystalline structures are cpx (clinopyroxene), β (β-phase), γ (γ-spinel), mw (magnesiowüstite), st (stishovite), pv (perovskite) (after Jeanloz and Thompson 1983).

($Mg_4Si_4O_{12}$). The aluminous garnet *pyrope* has a formula ($Mg_3Al_2Si_3O_{12}$) resulting from the replacement of one Si^{4+} ion by one Al^{3+} ion and compensating for the charge imbalance by replacing one Mg^{2+} by one Al^{3+}. At upper-mantle pressures, pyrope is soluble in the silicate garnet, yielding the aluminous silicate garnet *majorite,* first identified in shocked meteorites and later synthesized by Ringwood and Major (1971). Alumina widens the garnet stability field (Fig. 7.11). At pressures above about 200 kbar and temperatures lower than about 2000 K, $MgSiO_3$ garnet transforms into a phase with *ilmenite* structure (Liu 1976), which can be described as a corundum (Al_2O_3) structure with Mg^{2+} and Si^{4+} located in

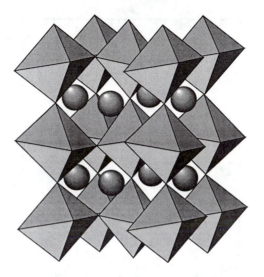

Figure 7.13. Structure of $MgSiO_3$ perovskite. The Mg cations are in dodecahedral sites between tilted corner-sharing SiO_6 octahedra. (The oxygen ions at the corners and the silicon ions at the centers of octahedra are not represented.)

an ordered alternate fashion in the 6-coordinated Al^{3+} sites. At lower-mantle pressures, the silicate ilmenite transforms into a phase with *perovskite* structure (Fig. 7.13), which can be described as a 3-D framework of corner-sharing SiO_6 octahedra, with Mg^{2+} in the dodecahedral sites. The slope of the phase boundary, in the range 1000–600°C, for the ilmenite–perovskite transition is negative, and is given (Ito and Takahashi 1989) by the relation P (GPa) $= 26.8 - 0.0025\ T$ (°C).

The existence of the perovskite phase was experimentally established by Liu, who synthesized it in the diamond–anvil cell, first from pyrope (Liu 1974), and then by decomposition of Mg_2SiO_4 at high pressure.

(MgFe)SiO_3 perovskite is probably the most abundant mineral in the Earth, since it constitutes possibly more than 80 vol% of the lower mantle. It is orthorhombic, at least up to 1 Mbar at room temperature, with a structure derived from the ideal cubic perovskite by tilting of the octahedra (Fig. 7.13) ($GdFeO_3$ distortion, see Muller and Roy 1974). Its principal physical properties, insofar as they have been experimentally determined, are listed in Table 7.2. The density and the equation of state parameters K_0 and K_0' are reasonably well known, having been determined by several investigators. The values of the thermal expansion coefficient and melting temperature, however, are still unconfirmed.

Table 7.2. *Physical properties (experimental) of MgSiO₃ perovskite*

Physical property		Experimental value	Reference
Cell dimensions (Å)		$a = 4.7787$	Ito and Matsui (1978)
		$b = 4.9313$	"
		$c = 6.9083$	"
Specific mass (g/cm³)		4.108	"
Mean atomic mass (g/atom)		20.08	"
Molar volume (cm³/mol)		24.4426	"
K_0 (GPa)	(Hill)	246.4	Yeganeh-Haeri et al. (1989)
K_0'		3.9	Knittle et al. (1987)
μ_0 (GPa)	(Hill)	184.2	Yeganeh-Haeri et al. (1989)
v_P (km/s)		10.94	"
v_S (km/s)		6.69	"
γ_{th}	(spectro)	1.9	Williams et al. (1987)
T_m (K)	(at 22 GPa)	3000	Heinz and Jeanloz (1987)
α (K⁻¹)	(298–840 K)	4×10^{-5}	Knittle and Jeanloz (1986)
	(298–341 K)	2.2×10^{-5}	Ross and Hazen (1989)

Phases with perovskite structure have also been found at high pressures for $CaSiO_3$ (Liu and Ringwood 1975), diopside $CaMgSi_2O_6$, and diopside–jadeite solid solutions ($CaMgSi_2O_6$–$NaAlSi_2O_6$) (Liu 1987). There is, however, no high-pressure phase with perovskite structure for the composition $FeSiO_3$, which decomposes into FeO and stishovite SiO_2 (Fig. 7.12).

iii. The phase transitions of Mg₂SiO₄

Olivine $(Mg_{1-x}Fe_x)_2SiO_4$ is an important, possibly dominant, mineral of the upper mantle (with $x \cong 0.1$). The magnesian end member Mg_2SiO_4, *forsterite,* forms a continuous series of solid solutions with *fayalite* Fe_2SiO_4. The orthorhombic olivine structure can be described as a slightly distorted hexagonal close-packed sublattice of oxygen ions, with the silicon ions occupying one-eighth of the tetrahedral sites and forming isolated SiO_4 tetrahedra, and the Mg or Fe ions occupying one-half of the octahedral sites. Alternatively, the structure can be considered as that of an intermetallic compound Mg_2Si (with the Ni_2In structure), stuffed with oxygen ions (O'Keeffe and Hyde 1981).

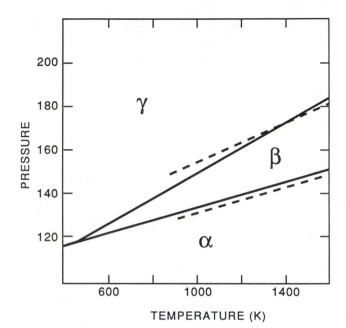

Figure 7.14. *P, T* phase diagram for the Mg₂SiO₄ polymorphs (α: olivine, β: modified spinel phase, γ: spinel). Solid lines from Akaogi et al. (1984), dashed lines
from Suito (1977) (after Akaogi et al. 1984).

High-pressure phases of the ferromagnesian olivines were first synthesized by Ringwood and Major (1966, 1970) and Akimoto and Fujisawa
(1966). Fayalite directly transforms into the spinel structure, still with
isolated SiO_4 tetrahedra, but with a face-centered cubic packing of oxygens. Forsterite and Mg-rich olivines first transform to an orthorhombic
phase with face-centered cubic packing of oxygens, where the SiO_4 tetrahedra are linked in pairs by a corner. This phase is called β-phase (α is olivine and γ is spinel) or "modified spinel." The *P–T* phase diagram of the
Mg_2SiO_4 polymorphs, constrained by thermochemical data (Akaogi et al.
1984) is given in Figure 7.14 and the pseudobinary diagram of Mg_2SiO_4–
Fe_2SiO_4 (Akaogi, Ito, and Navrotsky 1989) is given in Figure 7.15 (see
also Katsura and Ito 1989).

Natural high-pressure spinel phase or *ringwoodite,* was found in shocked
chondritic meteorites (Binns, Davis, and Reed 1969) and unambiguously identified by transmission electron microscopy by Putnis and Price
(1979) and Poirier and Madon (1979). The high-pressure β-phase (*wadsleyite*) was also found in shocked chondrites (Price et al. 1983; Madon
and Poirier 1983).

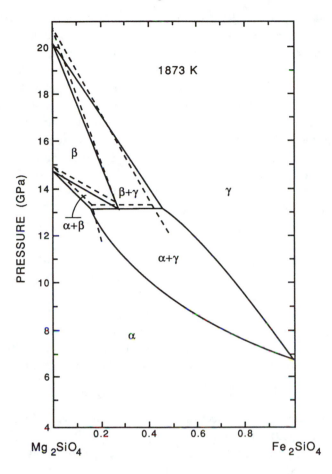

Figure 7.15. Isothermal ($T = 1600°\text{C}$) calculated phase diagram for the Mg_2SiO_4–Fe_2SiO_4 system. The crystalline structures are α (olivine), β (β-phase), γ (γ-spinel). The boundaries experimentally determined by Katsura and Ito (1989) are shown by dashed curves (after Akaogi et al. 1989).

The mechanism of the olivine–spinel transformation in silicates and analogue germanates has been much investigated in recent years. Since the oxygen sublattice goes from hexagonal close-packed in olivine to face-centered cubic in spinel, Poirier (1981a, b) proposed that by analogy with other hcp–fcc transformations, olivine could transform to spinel by shear restacking of oxygen ions due to invasion of the grains by stacking faults, the cations falling simultaneously into their new sites by "synchroshear." This "martensitic" mechanism was indeed found to be operative in experiments conducted in diamond–anvil cells (Lacam, Madon, and Poirier

1980; Boland and Liu 1983; Furnish and Bassett 1983), although there exists an intermediate stage, where the cations are disordered, thus ruling out synchroshear (Furnish and Bassett 1983). However, in experiments conducted in large-volume apparatus, the transformation was found to take place by nucleation and growth (Vaughan, Green, and Coe 1982; Boland and Liebermann 1983) with no cation disordering (Yagi et al. 1987), although disordering was found in experiments conducted in a belt-type apparatus (Lauterjung and Will 1986). It now seems clear that the transformation occurs by shear restacking of the oxygen ions only for experiments conducted under large shear stresses and/or with a large driving force (far from the Clapeyron) (Burnley and Green 1989), as is always the case for the diamond–anvil cell.

The seismic discontinuity at 670 km has been, for a long time, attributed to the transformation of $(Mg, Fe)_2SiO_4$ to denser post-spinel phases. It was first believed that spinel transformed to a mixture of magnesiowüstite $(Mg, Fe)O$ and stishovite, the high-pressure phase of quartz, where silicon is in octahedra SiO_6 ("mixed oxide" lower-mantle models). Ming and Bassett (1975) indeed found that the X-ray diffraction patterns of the disproportionated phase were consistent with the mixed oxide models. Liu (1975), however, conclusively showed that the post-spinel phases were a mixture of magnesiowüstite and the perovskite phase $(Mg, Fe)SiO_3$:

$$(Mg, Fe)_2SiO_4 \rightarrow (Mg, Fe)O + (Mg, Fe)SiO_3$$

Observations in analytical electron microscopy (Guyot et al. 1988; Madon, Guyot et al. 1989) give direct evidence of the validity of this disproportionation reaction. At high iron contents, however (higher than those currently admitted for the lower mantle), the perovskite is not stable and the disproportionation indeed leads to a mixture of magnesiowüstite and stishovite.

The pressure interval within which the decomposition takes place ("sharpness" of the transition) has long been a subject of controversy, bearing on the problem of the composition of the lower mantle (e.g., Lees et al. 1983) (see Chapter 10). Recent results (Ito and Takahashi 1989) show that magnesian spinel with less than 26 at% Fe dissociates within a very narrow pressure interval (1.5 kbar at 1600°C), thus buttressing the view that the transition is quite sharp (Fig. 7.16). The negative Clapeyron is given by P (GPa) $= 27.6 - 0.0028\ T$ (°C).

Iron does not disproportionate equally between magnesiowüstite and perovskite and goes preferentially into magnesiowüstite (Bell et al. 1979; Ito, Takahashi, and Matsui 1984). The partition coefficient of iron can be written

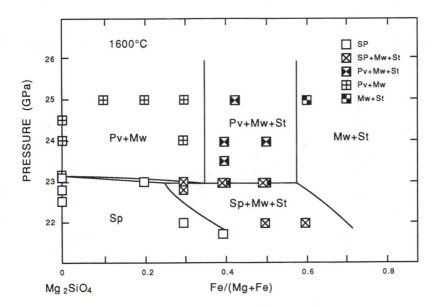

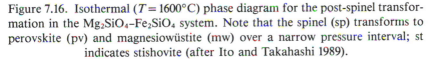

Figure 7.16. Isothermal ($T = 1600°C$) phase diagram for the post-spinel transformation in the Mg_2SiO_4–Fe_2SiO_4 system. Note that the spinel (sp) transforms to perovskite (pv) and magnesiowüstite (mw) over a narrow pressure interval; st indicates stishovite (after Ito and Takahashi 1989).

$$K = (x_{Fe}/x_{Mg})^{mw}/(x_{Fe}/x_{Mg})^{pv}$$

where x is the Fe or Mg content of the phases, given in at%. Guyot et al. (1988) have measured the value of K as a function of pressure by analytical transmission electron microscopy and found that it decreases with increasing pressure and remains constant ($K \cong 3.5$) at pressures above 400 kbar (Fig. 7.17). For the starting material used, it corresponds to the reaction:

$$(Mg_{0.83}, Fe_{0.17})_2SiO_4 \rightarrow (Mg_{0.91}, Fe_{0.09})SiO_3 + (Mg_{0.74}, Fe_{0.26})O$$

The structure of the disproportionated mixture of magnesiowüstite and perovskite was investigated as a function of T and P by transmission electron microscopy of the metastable quenched samples recovered from the diamond–anvil cell (Poirier et al. 1986; Madon, Guyot et al. 1989).

In this sketchy overview of the vast and developing field of the phase transitions that may occur in the mantle, we have left aside the relatively minor elements calcium and aluminum, not because their role is unimportant but because it is still largely unknown. Calcium and aluminum may enter into solution in the major ferromagnesian phases and modify phase diagram (see, e.g., Fig. 6.3); they also may form separate aluminocalcic

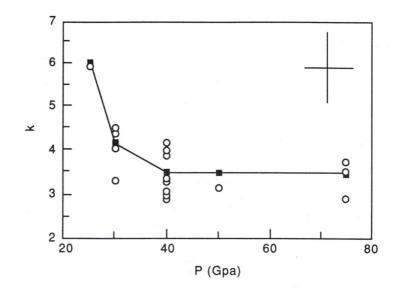

Figure 7.17. Variation with pressure of the partition coefficient of iron between perovskite and magnesiowüstite. The error bars on the measurements are shown in the upper right-hand corner (after Guyot et al. 1988a).

phases with structures still being investigated (Madon, Castex, and Peyronneau 1989).

7.4.2 Mantle and core models

i. Constraints and trade-offs

The compositional Earth models are usually patterned after the description of rocks by petrographers: For each region of the mantle, one must know the "norm," that is, the chemical composition expressed in percent by weight of oxides and the "mode," that is, the proportions of the various constituent minerals.

Compositional models are bound by different types of constraints.

They are required to agree as much as possible with the seismological models, that is, they must account for the velocity and density profiles. This of course implies that the *P–T* phase diagrams for the candidate compositions are known and that the equations of state, elastic moduli, and thermal expansion coefficient, as well as their temperature and pressure derivatives, are known for the relevant minerals. A geotherm must also be chosen.

The starting normative composition is usually chosen on the basis of assumptions about the primitive bulk Earth composition and the chemical evolution of the Earth.

Compositional models are often more or less openly tailored to fit other geophysical requirements: for example, whole mantle convection (hence chemically homogeneous mantle) or two-layer convection (hence the possibility for the upper and lower mantle to be chemically different). Note that this also conditions the choice of the geotherm.

The construction of a compositional model usually follows one of two lines.

i. The actual density and elastic moduli (often only the bulk modulus) at various depths are adiabatically extrapolated to zero pressure and brought down from the final temperature to room temperature. A mineral assemblage is devised so that its density and aggregate elastic properties fit the decompressed material.

ii. A mineral assemblage at room temperature and ambient pressure is devised so that after heating and adiabatically compressing it, its density and seismic velocities (often only the bulk velocity) fit the seismological model.

In both cases, the elastic moduli of the high-pressure phases and their pressure and temperature derivatives, either have been measured at ambient pressure in the metastable state, or, at any rate, have been estimated from elastic systematics. The thermal expansion coefficients often are estimated.

Most of the discrepancies between contending compositional Earth models come from one of two sources.

i. The models use different assumptions as to the primitive bulk Earth composition and its evolution. It is generally assumed that the primitive composition of the bulk Earth (and terrestrial planets) is that of the devolatilized solar nebula (Hart and Zindler 1986). Now, the problem is to decide what the composition of the original solar nebula was and to what degree it has lost volatile elements. It is currently thought that the solar abundance of refractory elements is well reflected in the composition of the $C1$ chondrites, meteorites that have not been differentiated and are considered as samples of the primitive nebular material (Anders and Grevesse 1989). The composition of the primitive upper mantle, however, derived from the study of mantle peridotites, corresponds to a greater Mg/Si ratio than that of the chondrites; it has been recently proposed that the Mg/Si ratio of the terrestrial planets is more representative of the solar nebula value than that of the $C1$ chondrites (Ringwood 1989).

If one believes that the bulk Earth is nevertheless chondritic, one is drawn to the conclusion that the missing silicon is hidden away in the lower mantle and core, and the resulting compositional models are called *chondritic Earth models* (e.g., Liu 1982; D. Anderson 1984; Anderson and Bass 1986). In these models, the lower mantle is more silica-rich than the upper mantle and can be composed almost entirely of $(Mg, Fe)SiO_3$ perovskite, thus leading to a two-layer mantle convection pattern.

One can also solve the case of the missing silicon by assuming that this element, less refractory than calcium and aluminum, was partly volatilized away. The mantle then need not be chemically heterogeneous and whole-mantle convection is not precluded (although not required). The most popular model of homogeneous mantle is the *pyrolite model* (see Ringwood 1979): The bulk composition of the mantle is that of "pyrolite," a nonspecific olivine-pyroxene rock, capable of yielding a basaltic magma and a peridotite residue upon partial melting in the uppermost mantle. Pyrolite is a fictitious rock in that it is defined by its chemical composition only and not by its mineralogy.

Recently, D. Anderson (1989), using new estimates of the solar composition, proposed that the Sun should be richer in iron and calcium than the $C1$ chondrites. If the bulk Earth composition is solar, it then should be richer in iron and calcium than the chondritic model. The composition of the mantle corresponding to the various hypotheses is given in Table 7.3.

ii. The other source of discrepancy lies in the choice of the elastic and thermal parameters of the high-pressure candidate minerals. Most of them are, if not unknown, at least known with a high degree of uncertainty, stemming either from the inherent experimental errors and/or from the fact that we must often rely on only one measurement. Also, the elastic parameters are sensitive to the assumed iron content. As a consequence, many trade-offs are possible and various mineral assemblages may be fitted to the velocity and density profiles (themselves only known within the resolution of the seismological methods) by choosing an appropriate geotherm and not unlikely combinations of elastic moduli, thermal expansion coefficient, and P–T derivatives (see Jackson 1983). It might be submitted that at the present time, there is no compelling evidence in favor of any one of the contending models, such evidence as is usually presented for one model never really ruling out other models. There is a

Table 7.3. *Composition models for the Earth's mantle, in wt%,
for the five major oxides*

Oxide	Pyrolite[a]	Chondritic[b]	Chondritic[c]	Chondritic[d]	Solar[e]
SiO_2	45.0	50.8	49.52	45.96	45
MgO	38.8	36.6	35.68	37.78	32.7
FeO	7.6	6.08	7.14	7.54	15.7
Al_2O_3	4.4	3.67	3.56	4.06	3.2
CaO	3.4	2.89	2.82	3.21	3.4

[a] Jackson (1983)
[b] Anderson and Bass (1986)
[c] Hart and Zindler (1986)
[d] Hart and Zindler (1986) devolatilized chondritic (LOSIMAG)
[e] Anderson (1989)

pressing need for accurate laboratory measurements of elastic and thermal parameters of high-pressure mantle phases and of their pressure and temperature derivatives.

ii. Mantle models

We have some direct knowledge of the composition of the uppermost mantle from the produce of its partial melting, basalts, and from the peridotites found in ophiolitic complexes, massifs, and xenoliths brought up from as deep as 200 or 250 km by basalts or kimberlites. Ringwood (1979) draws the conclusion that the uppermost mantle is composed of residual peridotite strongly depleted in the low melting-point minerals that went into basalt during partial melting. A typical peridotite is composed of mostly olivine and orthopyroxene (enstatite) with some calcic clinopyroxene (diopside) and an aluminous phase (plagioclase, spinel, or pyrope garnet, in that order with increasing depth). Below the lithosphere, there must exist a primitive source material: pyrolite, defined as we have already seen by its capacity to produce basalt and residual peridotite. The phase transitions of the relevant minerals, described previously, are compatible with the following model.

At about 350 km, pyroxene and aluminous garnet enter into solid solution, giving majorite with garnet structure, and at 400 km, olivine goes to β-phase; then, in the transition zone, β-phase goes to γ-spinel, garnet goes to ilmenite, and the $CaSiO_3$ component goes to calcic perovskite;

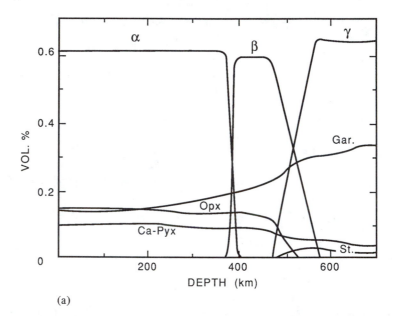

(a)

Figure 7.18a. The calculated volume fraction of mineral phases for a pyrolite mantle composition as a function of depth (after Weidner 1986).

finally, at 650 km, $(Mg, Fe)SiO_3$ disproportionates to perovskite and magnesiowüstite. Ringwood (1975) found that these transitions provided a satisfactory explanation of the position and magnitude of the seismic discontinuities and density jumps and that there was no need for changes in chemical composition and, in particular, no need for an enrichment in iron of the lower mantle. The 650-km seismic discontinuity is then thought to be due only to phase transitions, but it is seismically marked by the thin subducted oceanic lithosphere trapped at the interface between upper and lower mantle (Ringwood and Irifune 1988).

D. Anderson (1984), Anderson and Bass (1986), and Duffy and Anderson (1989) account for the seismic velocity profiles by mineralogical assemblages consistent with a chondritic mantle having melted (magma ocean) and differentiated at an early stage of the Earth's history: Low-density olivine crystallizes first and concentrates into a peridotite uppermost mantle; the residual fluid freezes to a clinopyroxene garnet-rich assemblage with less than 50 percent olivine, termed *piclogite,* which would constitute the transition zone. The lower-mantle velocities are consistent with a silica-rich composition of pure perovskite, as also proposed by Liu (1979).

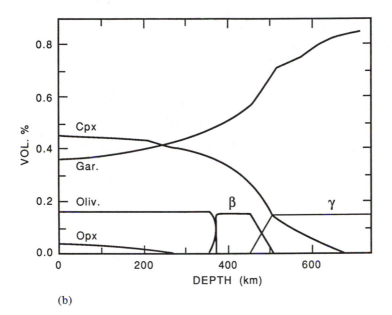

(b)

Figure 7.18b. The calculated volume fraction of mineral phases for a piclogite mantle composition as a function of depth (after Weidner 1986).

In the recent D. Anderson (1989) model, the lower mantle is also richer in iron and there is more diopside in the transition zone.

The two major classes of upper-mantle models are therefore (see Fig. 7.18 and Table 7.4):

the "pyrolite" mantle models, with no chemical difference between upper and lower mantle; and

the "piclogite" mantle models, with a more silica- (and iron-) rich lower mantle.

Jackson (1983) showed that due to the trade-offs among composition, temperature, and physical properties, pyrolite and chondritic models could be equally plausible. Experimental investigation of the phase relations of the $MgO–FeO–SiO_2$ system (Ito et al. 1984) led the authors to the conclusion that a pyrolite lower mantle satisfied the density and bulk modulus constraints, although a pure perovskite composition or a composition close to that of E enstatite chondrites could not be ruled out. Poirier (1988b) investigated the variation of Poisson's ratio of the perovskite– magnesiowüstite assemblage with various perovskite and iron contents

Table 7.4. *Upper mantle models*

Oxide	Pyrolite	Piclogite
Norm		
MgO	40.3	21.0
FeO	7.9	5.7
CaO	3.0	7.0
SiO_2	45.2	48.9
Al_2O_3	3.5	14.4
Na_2O	0.0	3.0
Mode		
Olivine	61	16
Orthopyroxene	15	3
Clinopyroxene	10	16
Garnet	14	36

Note: Norm in wt% oxides, mode in vol%
minerals.
Source: Weidner (1986).

and found that the Poisson's ratio and the seismic parameter of the lower
mantle was satisfied both by a pyrolite and a chondritic model. Weidner
(1986) investigated the agreement between the upper-mantle models and
the seismological data. He found that the phase transitions in the transition zone alone could satisfy the data without resorting to mineralogical
stratification as in "piclogite" models, but, however, they cannot be cate-
gorically ruled out.

In conclusion, it is difficult to eliminate one of the models on the grounds
that it does not fit the seismological data and it is possible to say with
Weidner (1986) that "within the uncertainty of the current data base, we
need look no further than the pyrolite model to find a chemical composition compatible with the data for the upper mantle"; as for the lower
mantle it is even less well constrained. Note again that a homogeneous
mantle does *not* necessarily imply whole-mantle convection.

iii. The core–mantle boundary

The core–mantle boundary (CMB) is possibly the major discontinuity in
the Earth, with a density contrast of 4.4 g/cm^3 between the core and the
mantle (compared to 2.7 g/cm^3 between the crust and the atmosphere)

and a viscosity contrast of the order of 10^{21} Poises. It is the seat of energetic exchanges and couplings between core and mantle that may be of extreme geodynamic importance and it is becoming the object of a sustained interest. The reader is referred to recent reviews by Young and Lay (1987) and Lay (1989), as well as to a special issue of *Geophysical Research Letters* (volume 13, n. 13, 1986).

Global seismological Earth models exhibit a zone of reduced or even negative velocity gradient (Lay 1989), extending 200 to 300 km above the CMB, called the D'' layer. Since, from most estimates, there is a finite heat flux from the core into the mantle, it was natural to think of the D'' zone as just a thermal boundary layer. However, recent seismological investigations, using travel time residuals of body waves reflected from the core or transmitted through it, show that the structure of the D'' layer is rather complicated. The CMB may be bumpy, with a "topography" of several kilometers amplitude at the scale of thousands of kilometers (Morelli and Dziewonski 1987) and there also is evidence of lateral heterogeneities of velocity, which can be interpreted as due to rafts of denser material, unevenly distributed at the CMB, much like the continents at the top of the mantle (Dornboos, Spiliopoulos, and Stacey 1986). The analysis of the correlations between the geoid and the surface topography shows that the amplitude of the topography dynamically maintained by convection at the core–mantle boundary should be about 3 km (Hager et al. 1985).

Numerical simulations lead to a picture of the D'' zone as an unstable thermal boundary layer over the depth of which the temperature-dependent viscosity can decrease by several orders of magnitude, inducing the rise of convective instabilities, or plumes (Loper 1984; Zharkov, Karpov, and Leontieff 1985; Olson et al. 1987).

Chemically denser material (dregs) may lie at the bottom of the mantle, in an uneven layer of variable thickness, interacting with the convection and modulating the heat flow from the core (Davies and Gurnis 1986). The nature of the dense material is still as speculative as its existence; it seems, however, certain that the molten iron alloy of the core can react with the silicate perovskite and magnesiowüstite (Urakawa, Kato, and Kumazawa 1987; Williams, Knittle, and Jeanloz 1987; Guyot et al. 1988; Knittle and Jeanloz 1989). If, as convincingly argued by Stevenson (1981), the core is not in equilibrium with the mantle, the core–mantle boundary must then be a chemically active and probably heterogeneous zone.

The current image of the D'' layer is therefore that of a chemical boundary layer imbedded in a thermal boundary layer (Lay 1989).

iv. The core

The physical state and composition of the Earth's core have been a matter of considerable controversy up to recent times (see Brush 1979, 1982) and there still is some debate as to exactly what elements it contains in addition to iron. That the core is made of iron is consistent with the large increase in density at the core–mantle boundary and with the idea that iron meteorites constitute the cores of small differentiated planetary bodies.

The view had been entertained (Ramsey 1949) that the core, like the mantle, was composed of silicate phases, compressed to a very high density and having undergone a transition to the metallic state. Birch (1952, 1961b, 1963), using the bulk velocity–density systematics he had established for a number of elements and the Hugoniot compression curves experimentally obtained for iron, conclusively showed that the core was indeed iron (see Sec. 4.5.2 and Fig. 4.4). His principal argument was that silicates, with a mean atomic mass close to 20, could never achieve the density of the core given by seismological models, unless under pressures much larger than the pressure at the center of the Earth (or that the bulk sound velocity of silicates at core pressure densities would be much too high); iron, on the contrary, gave good agreement with the density and seismic velocities, if alloyed with a small quantity of light elements. Even though Ramsey's theory is revived from time to time, it is clearly not tenable (O. Anderson 1985).

By analogy with iron meteorites and from cosmic abundances considerations, it is reasonable to assume that there is some nickel in the core. If the Earth is of cosmic composition, there should be about 4 wt% nickel in the Earth's core (Brett 1976).

A core of pure iron (and a fortiori a core containing some nickel) would have a density about 10 percent higher than that deduced from seismological observations (Birch 1952). The density may be adjusted by assuming that iron-nickel is alloyed with some proportion of light elements. What the light elements are is an object of debate (see Jacobs 1986; Stevenson 1981 for reviews). It must be noted that the arguments for and against a given element often rely on assumptions about the bulk Earth composition and the mode of formation of the core.

Silicon may well be present in the core, although there are some difficulties with the models that claim it is the best candidate (see Stevenson 1981).

Oxygen is little soluble in iron near its liquidus at atmospheric pressure, but its solubility increases with temperature and probably with pressure. Ringwood (1977, 1979) estimates the oxygen content of the core to be 10 ± 4 wt% (44 ± 16 wt% FeO). The probability for oxygen to be present in the core is enhanced by recent experimental results showing that FeO, immiscible with Fe at ambient pressure, becomes metallic above 700 kbar and can thus be incorporated in the iron core (Knittle and Jeanloz 1986). The metallization may be due to a Mott transition (Sherman 1989).

Sulfur also remains a good candidate since it easily partitions into iron and can form a low melting-point eutectic with it; 8 to 10 wt% S would be enough to account for the core density (Stevenson 1981).

Potassium might be present in the core, especially if there is sulfur. The radioactive decay of ^{40}K might then contribute to the heat flow from the core (Stacey 1972).

Hydrogen has recently been found to be highly soluble in iron at high pressure and to form hydrides. It has been suggested that hydrogen might be one of the light elements present in the core (Suzuki, Akimoto, and Fukai 1984; Fukai and Suzuki 1986).

Of course, as pointed out by Stevenson (1981), there is no reason to believe that the core is a particularly "clean" system and that there is only one light element present in it. The light elements, whatever they are, partition in the liquid phase during crystallization of the iron alloy at the inner core boundary; the lighter liquid can then rise and the ensuing "solutal convection" is thought to be essential to the generation of the magnetic field.

The inner core freezes out as essentially pure iron, possibly, but not certainly, in the ϵ-phase (see Sec. 5.5). The density and seismic velocities of pure iron generally agree with those of the inner core. However, Jephcoat and Olson (1987) find that a significant amount of light component should be incorporated in the inner core.

The study of the Earth's deep interior is being actively pursued and it is not unlikely, indeed it is to be hoped, that by the time the present book is off the presses, part of this last chapter will be in need of updating.

Appendix

Table A.1. *PREM model (1s) for the mantle and core*

z	r	P	ρ	v_P	v_S	Φ	K	μ	ν	g
24.4	6346	6	3.38	8.11	4.49	38.9	1315	682	0.28	984
40	6331	11.2	3.38	8.11	4.48	38.8	1311	680	0.28	984
60	6311	17.9	3.38	8.09	4.48	38.7	1307	677	0.28	985
80	6291	24.5	3.37	8.08	4.47	38.6	1303	674	0.28	986
115	6256	36.2	3.37	8.03	4.44	38.2	1287	665	0.28	988
185	6186	59.4	3.36	8.01	4.43	38.0	1278	660	0.28	989
220	6151	71.1	3.36	7.99	4.42	37.8	1270	656	0.28	990
220	6151	71.1	3.44	8.56	4.64	44.5	1529	741	0.29	990
265	6106	86.5	3.42	8.65	4.68	45.6	1579	757	0.29	992
310	6061	102	3.49	8.73	4.71	46.7	1630	773	0.30	994
355	6016	118	3.52	8.81	4.74	47.8	1682	790	0.30	995
400	5971	134	3.54	8.91	4.77	49.0	1735	806	0.30	997
400	5971	134	3.72	9.13	4.93	51.0	1899	906	0.29	997
450	5921	152	3.79	9.39	5.08	53.8	2037	977	0.29	998
500	5871	171	3.85	9.65	5.22	56.7	2181	1051	0.29	999
550	5821	191	3.91	9.90	5.37	59.6	2332	1128	0.29	1000
600	5771	210	3.98	10.16	5.51	62.6	2489	1210	0.29	1000
635	5736	224	3.98	10.21	5.54	63.3	2523	1224	0.29	1001
670	5701	238	3.99	10.27	5.57	64.0	2556	1239	0.29	1001
670	5701	238	4.38	10.75	5.95	68.5	2999	1548	0.28	1001
721	5650	261	4.41	10.91	6.09	69.5	3067	1639	0.27	1001
771	5600	283	4.44	11.07	6.24	70.5	3133	1730	0.27	1000
871	5500	328	4.50	11.24	6.31	73.3	3303	1794	0.27	999
971	5400	373	4.56	11.41	6.38	76.1	3471	1856	0.27	997
1071	5300	419	4.62	11.58	6.44	78.7	3638	1918	0.28	996
1171	5200	465	4.68	11.73	6.50	81.3	3803	1979	0.28	995
1271	5100	512	4.73	11.88	6.56	83.8	3966	2039	0.28	994
1371	5000	559	4.79	12.02	6.62	86.2	4128	2098	0.28	993
1471	4900	607	4.84	12.16	6.67	88.5	4288	2157	0.28	993
1571	4800	655	4.90	12.29	6.72	90.8	4448	2215	0.29	993
1671	4700	704	4.95	12.42	6.77	93.1	4607	2273	0.29	994
1771	4600	754	5.00	12.54	6.83	95.3	4766	2331	0.29	995
1871	4500	804	5.05	12.67	6.87	97.4	4925	2388	0.29	996
1971	4400	854	5.11	12.78	6.92	99.6	5085	2445	0.29	999
2071	4300	906	5.16	12.90	6.97	101.7	5246	2502	0.29	1002
2171	4200	958	5.21	13.02	7.01	103.9	5409	2559	0.30	1005
2271	4100	1010	5.26	13.13	7.06	106.0	5575	2617	0.30	1010
2371	4000	1064	5.31	13.25	7.10	108.2	5744	2675	0.30	1016
2471	3900	1118	5.36	13.36	7.14	110.5	5917	2734	0.30	1023

Table A.1 *(cont.)*

z	r	P	ρ	v_P	v_S	Φ	K	μ	ν	g
2571	3800	1173	5.41	13.48	7.19	112.7	6095	2794	0.30	1031
2671	3700	1230	5.46	13.60	7.23	115.1	6279	2855	0.30	1041
2771	3600	1287	5.51	13.67	7.27	117.0	6440	2907	0.30	1052
2871	3500	1346	5.56	13.71	7.26	117.6	6537	2933	0.30	1065
2891	3480	13581	5.57	13.72	7.26	117.8	6556	2938	0.31	1068
2891	3480	13581	9.90	8.06	0	65.0	6441	0	0.5	1068
2971	3400	1442	10.02	8.19	0	67.2	6743	0	0.5	1051
3071	3300	1547	10.18	8.36	0	69.9	7116	0	0.5	1028
3171	3200	1651	10.33	8.51	0	72.5	7484	0	0.5	1005
3271	3100	1754	10.47	8.66	0	75.0	7846	0	0.5	981
3371	3000	1856	10.60	8.80	0	77.4	8202	0	0.5	956
3471	2900	1957	10.73	8.93	0	79.7	8550	0	0.5	930
3571	2800	2056	10.85	9.05	0	81.9	8889	0	0.5	904
3671	2700	2153	10.97	9.17	0	84.0	9220	0	0.5	877
3771	2600	2248	11.08	9.28	0	86.1	9542	0	0.5	850
3871	2500	2342	11.19	9.38	0	88.1	9855	0	0.5	822
3971	2400	2342	11.29	9.48	0	90.0	10158	0	0.5	794
4071	2300	2521	11.39	9.58	0	91.8	10451	0	0.5	766
4171	2200	2607	11.48	9.67	0	93.5	10735	0	0.5	736
4271	2100	2690	11.57	9.75	0	95.1	11009	0	0.5	707
4371	2000	2770	11.65	9.83	0	96.7	11273	0	0.5	677
4471	1900	2848	11.73	9.91	0	98.3	11529	0	0.5	647
4571	1800	2922	11.81	9.99	0	99.1	11775	0	0.5	617
4671	1700	2993	11.88	10.05	0	101.1	12013	0	0.5	586
4771	1600	3061	11.95	10.12	0	102.5	12242	0	0.5	555
4871	1500	3126	12.01	10.19	0	103.8	12464	0	0.5	524
4971	1400	3187	12.07	10.25	0	105.1	12679	0	0.5	494
5071	1300	3245	12.13	10.31	0	106.3	12888	0	0.5	464
5150	1221	3289	12.17	10.36	0	107.2	13047	0	0.5	440
5150	1221	3289	12.76	11.02	3.50	105.3	13434	1567	0.44	440
5171	1200	3300	12.77	11.03	3.51	105.4	13462	1574	0.44	432
5271	1100	3354	12.83	11.07	3.54	106.0	13586	1603	0.44	397
5371	1000	3402	12.87	11.11	3.56	106.5	13701	1630	0.44	362
5471	900	3447	12.91	11.14	3.58	106.9	13805	1654	0.44	326
5571	800	3487	12.95	11.16	3.60	107.3	13898	1676	0.44	291
5671	700	3522	12.98	11.18	3.61	107.7	13981	1696	0.44	255
5771	600	3553	13.01	11.21	3.63	108.2	14053	1713	0.44	217
5871	500	3579	13.03	11.22	3.64	108.3	14114	1727	0.44	182
5971	400	3600	13.05	11.24	3.65	108.5	14164	1739	0.44	146
6071	300	3617	13.07	11.25	3.66	108.7	14203	1749	0.44	110
6171	200	3629	13.08	11.26	3.66	108.8	14231	1755	0.44	73
6271	100	3636	13.09	11.26	3.67	108.9	14248	1759	0.44	37
6371	0	3639	13.09	11.26	3.67	108.9	14253	1761	0.44	0

Note: z: depth, in km; r: radius, in km; P: pressure, in kbar; $ρ$: specific mass, in g/cm^3; v_P: P-wave velocity, in km/s; v_S: S-wave velocity, in km/s; $Φ$: seismic parameter, in km^{2}s^{-2}; K: bulk modulus, in kbar; $μ$: shear modulus, in kbar; $ν$: Poisson's ratio; g: acceleration of gravity, in cm s^{-2}.

References

Ahrens, T. J. (1971). Shock-wave equation of state of minerals. In *Scuola Int. di Fisica "E. Fermi": "Mantello e nucleo nella fisica planetaria,"* J. Coulomb and M. Caputo, eds. New York: Academic.

(1980). Dynamic compression of Earth materials. *Science* **207**, 1035–41.

(1987). Shock wave techniques for geophysics and planetary physics. In *Methods of experimental physics,* C. G. Sammis and T. L. Henyey, eds. New York: Academic, **24 A**, 185–235.

Ahrens, T. J.; Anderson, D. L.; and Ringwood, A. E. (1969). Equations of state and crystal structures of high-pressure phases of shocked silicates and oxides. *Rev. Geophys.* **7**, 667–707.

Akaogi, M.; Ito, E.; and Navrotsky, A. (1989). Olivine–modified spinel–spinel transitions in the system Mg_2SiO_4–Fe_2SiO_4: calorimetric measurements, thermochemical calculation, and geophysical application. *J. Geophys. Res.* **94**, 15671–85.

Akaogi, M.; Navrotsky, A.; Yagi, T.; and Akimoto, S. (1987). Pyroxene–garnet transformation: Thermochemistry and elasticity of garnet solid solutions and application to a pyrolite mantle. In *High-pressure research in mineral physics,* M. H. Manghnani and Y. Syono, eds. Washington, D.C.: American Geophysical Union, 251–60.

Akaogi, M.; Ross, N.; McMillan, P.; and Navrotsky, A. (1984). The Mg_2SiO_4 polymorphs (olivine, modified spinel and spinel) – Thermodynamic properties from oxide melt solution calorimetry, phase relations and models of lattice vibrations. *Amer. Mineral.* **69**, 499–512.

Akimoto, S. (1987). High-pressure research in geophysics: Past, present and future. In *High-pressure research in mineral physics,* M. H. Manghnani and Y. Syono, eds. Washington, D.C.: American Geophysical Union, 1–13.

Akimoto, S., and Fujisawa, H. (1965). Demonstration of the electrical conductivity jump produced by the olivine spinel transition. *J. Geophys. Res.* **70**, 443–9.

(1966). Olivine–spinel transition in the system Mg_2SiO_4–Fe_2SiO_4 at 800°C. *Earth Planet. Sci. Lett.* **1**, 237–40.

Akimoto, S.; Suzuki, T.; Yagi, T.; and Shimomura, O. (1987). Phase diagram of iron determined by high pressure/temperature X-ray diffraction using synchrotron radiation. In *High-pressure research in mineral physics,* M. H. Manghnani and Y. Syono, eds. Washington, D.C.: American Geophysical Union, 149–54.

239

Alers, G. A. (1965). Use of sound velocity measurements in determining the Debye temperature of solids. In *Physical acoustics, vol. III B.* W. P. Mason and R. N. Thurston, eds. New York: Academic.

Allègre, C. J. (1982). Chemical geodynamics. *Tectonophysics* **81**, 109–32.

(1987). Isotope geodynamics. *Earth Planet. Sci. Lett.* **86**, 175–203.

Allègre, C. J., and Turcotte, D. L. (1987). Geodynamic mixing in the mesosphere layer and the origin of oceanic islands. *Geophys. Res. Lett.* **12**, 207–10.

Anders, E., and Grevesse, N. (1989). Abundance of the elements: Meteoritic and solar. *Geochim. Cosmochim. Acta* **35**, 197–214.

Anderson, D. L. (1967). A seismic equation of state. *Geophys. J. R. Astr. Soc.* **13**, 9–30.

(1984). The Earth as a planet: Paradigms and paradoxes. *Science* **223**, 347–55.

(1987a). A seismic equation of state, II. Shear properties and thermodynamics of the lower mantle. *Phys. Earth Planet. Interiors* **45**, 307–23.

(1987b). Thermally induced phase changes, lateral heterogeneity of the mantle, continental roots and deep slab anomalies. *J. Geophys. Res.* **92**, 13968–80.

(1988). Temperature and pressure derivatives of elastic constants, with application to the mantle. *J. Geophys. Res.* **93**, 4668–700.

(1989). Composition of the Earth. *Science* **243**, 367–70.

Anderson, D. L., and Anderson, O. L. (1970). The bulk modulus–volume relationship for oxides. *J. Geophys. Res.* **75**, 3494–500.

Anderson, D. L., and Bass, J. D. (1986). Transition region of the Earth's upper mantle. *Nature* **320**, 321–8.

Anderson, D. L., and Kanamori, H. (1968). Shock-wave equations of state for rocks and minerals. *J. Geophys. Res.* **73**, 6477–502.

Anderson, O. L. (1963). A simplified method for calculating the Debye temperature from elastic constants. *J. Phys. Chem. Solids* **24**, 909–17.

(1979). Evidence supporting the approximation $\gamma\rho = \text{const}$ for the Grüneisen parameter of the Earth's lower mantle. *J. Geophys. Res.* **84**, 3537–42.

(1980). An experimental high-temperature thermal equation of state bypassing the Grüneisen parameter. *Phys. Earth Planet. Interiors* **22**, 173–83.

(1982). The Earth's core and the phase diagram of iron. *Phil. Trans. R. Soc. Lond.* **A306**, 21–35.

(1984). A universal thermal equation of state. *J. Geodynamics* **1**, 185–214.

(1985). Ramsey's silicate core revisited. *Nature* **314**, 407–8.

(1986). Properties of iron at the Earth's core conditions. *Geophys. J. R. Astr. Soc.* **84**, 561–79.

(1988). Simple solid-state equations for materials of terrestrial planet interiors. In *The physics of planets,* S. K. Runcorn, ed. New York: Wiley.

Anderson, O. L., and Nafe, J. E. (1965). The bulk modulus–volume relationship for oxide compounds and related geophysical problems. *J. Geophys. Res.* **70**, 3951–63.

Anderson, O. L., and Sumino, Y. (1980). The thermodynamic properties of the Earth's lower mantle. *Phys. Earth Planet. Interiors* **23**, 314–31.

Anderson, O. L., and Suzuki, I. (1983). Anharmonicity of three minerals at high temperature: forsterite, fayalite and periclase. *J. Geophys. Res.* **88**, 3549–56.

Anderson, P. W. (1984). *Basic notions of condensed matter physics,* pp. 49–69. London: Benjamin.

Andrade, E. N. da C. (1934). A theory of the viscosity of liquids. *Philos. Mag.* **17**, 497–732.

(1952). Viscosity of liquids. *Proc. R. Soc. London* **215A**, 36–43.

Animalu, A. O. E. (1977). *Intermediate quantum theory of crystalline solids.* Englewood Cliffs, N.J.: Prentice-Hall.

Appel, J. (1968). Polarons. *Sol. State Phys.* **21**, 193–391.

Austin, I. G., and Mott, N. F. (1969). Polarons in crystalline and noncrystalline materials. *Adv. Phys.* **18**, 41–102.

Babb, S. E. (1963a). Parameters in the Simon equation relating pressure and melting temperature. *Rev. Mod. Phys.* **35**, 400–13.

(1963b). Values of the Simon constants. *J. Chem. Phys.* **38**, 2743–9.

Balachandran, U.; Odekirk, B.; and Eror, N. G. (1982). Electrical conductivity of calcium titanate. *J. Solid State Chem.* **41**, 185–94.

Banks, B. E. C.; Damjanovic, V.; and Vernon, C. A. (1972). The so-called thermodynamic compensation law and thermal death. *Nature* **240**, 147–8.

Barin, I., and Knacke, O. (1973). *Thermochemical properties of inorganic substances.* Berlin: Springer.

Bassett, W. A. (1977). The diamond cell and the nature of the Earth's interior. *Ann. Rev. Earth Planet. Sci.* **7**, 357–84.

Bassett, W. A., and Huang, E. (1987). Mechanism of the body-centered cubic–hexagonal close-packed phase transition in iron. *Science* **238**, 780–3.

Battezzati, L., and Greer, A. L. (1989). The viscosity of liquid metals and alloys. *Acta Metall.* **37**, 1791–802.

Bell, P. M.; Mao, H. K.; and Xu, J. A. (1987). Error analysis in parameter-fitting in equations of state for mantle minerals. In *High-pressure research in mineral physics,* M. H. Manghnani and Y. Syono, eds. Washington, D.C.: American Geophysical Union, 447–54.

Bell, P. M.; Yagi, T.; and Mao, H. K. (1979). Iron–magnesium distribution coefficients between spinel $(Mg,Fe)_2SiO_4$, magnesiowüstite $(Mg,Fe)O$ and perovskite $(Mg,Fe)SiO_3$. *Carnegie Inst. Washington Yearbook* **78**, 618–21.

Bercovici, D.; Schubert, G.; and Glatzmaier, G. A. (1989). Three-dimensional spherical models of convection in the Earth's mantle. *Science* **244**, 950–5.

Berger, J., and Joigneau, S. (1959). Au sujet de la relation linéaire existant entre la vitesse matérielle et la vitesse de l'onde de choc se propageant dans un métal. *C. R. Acad. Sci. Paris* **249**, 2506–8.

Berman, R. (1976). *Thermal conduction in solids.* Oxford: Oxford University Press.

Binns, R. A.; Davis, R. J.; and Reed, S. J. B. (1969). Ringwoodite, natural Mg_2SiO_4 spinel in the Tenham meteorite. *Nature* **221**, 943–4.

Birch, F. (1938). The effect of pressure upon the elastic parameters of isotropic solids, according to Murnaghan's theory of finite strain. *J. Appl. Phys.* **9**, 279–88.

(1947). Finite elastic strain of cubic crystals. *Phys. Rev.* **71**, 809–924.

(1952). Elasticity and constitution of the Earth's interior. *J. Geophys. Res.* **57**, 227–86.

(1961a). The velocity of compressional waves in rocks to 10 kilobars, Part 2. *J. Geophys. Res.* **66**, 2199–224.

(1961b). Composition of the Earth's mantle. *Geophys. J. R. Astr. Soc.* **4**, 295–311.

(1963). Some geophysical applications of high-pressure research. In *Solids under pressure,* W. Paul and D. M. Warschauer, eds. New York: McGraw-Hill, 137–62.

(1964). Density and composition of mantle and core. *J. Geophys. Res.* **69**, 4377–88.

(1968). Thermal expansion at high pressures. *J. Geophys. Res.* **73**, 817–19.

(1972). The melting relations of iron and temperatures in the Earth's core. *Geophys. J. R. Astr. Soc.* **29**, 373–87.

Bocquet, J. L.; Brébec, G.; and Limoge, Y. (1983). Diffusion in metals and alloys. In *Physical metallurgy,* R. W. Cahn and P. Haasen, eds. New York: Elsevier, 386–475.

Boehler, R. (1982). Adiabats of quartz, coesite, olivine and magnesium oxide to 50 kbar and 100K, and the adiabatic gradient in the Earth's mantle. *J. Geophys. Res.* **87**, 5501–6.

(1986). The phase diagram of iron to 430 kbar. *Geophys. Res. Lett.* **13**, 1153–6.

Boehler, R., and Ramakrishnan, J. (1980). Experimental results on the pressure dependence of the Grüneisen parameter: a review. *J. Geophys. Res.* **85**, 6996–7002.

Boland, J. N., and Liebermann, R. C. (1983). Mechanism of the olivine to spinel phase transformation in Ni_2SiO_4. *Geophys. Res. Lett.* **10**, 87–90.

Boland, J. N., and Liu, L. G. (1983). Olivine to spinel transformation in Mg_2SiO_4 via faulted structures. *Nature* **303**, 233–5.

Bolt, B. A. (1982). *Inside the Earth.* San Francisco: Freeman.

Boon, M. R. (1973). Thermodynamic compensation rule. *Nature* **243**, 401.

Borg, R. J., and Dienes, G. J. (1988). *An introduction to solid state diffusion.* New York: Academic.

Born, M. (1939). Thermodynamics of crystals and melting. *J. Chem. Phys.* **7**, 591–603.

Boschi, E. (1974). Melting of iron. *Geophys. J. R. Astr. Soc.* **38**, 327–44.

Boschi, E., and Caputo, M. (1969). Equations of state at high pressure and the Earth's interior. *Riv. Nuovo Cimento* **1**, 441–513.

Bottinga, Y. (1985). On the isothermal compressibility of silicate liquids at high pressure. *Earth Planet. Sci. Lett.* **74**, 350–60.

Bowen, H. K.; Adler, D.; and Auker, B. H. (1975). Electrical and optical properties of FeO. *J. Solid State Chem.* **12**, 355–9.

Bradley, R. S.; Jamil, A. K.; and Munro, D. C. (1964). The electrical conductivity of olivine at high temperature and pressure. *Geochim. Cosmochim. Acta* **28**, 1669–78.

Bradley, R. S.; Milnes, G. J.; and Munro, D. C. (1973). The electrical conductivities at elevated temperatures and pressures of polycrystalline manganese, cobalt and nickel orthosilicates. *Geochim. Cosmochim. Acta* **37**, 2379–94.

Brennan, B. J., and Stacey, F. D. (1979). A thermodynamically based equation of state for the lower mantle. *J. Geophys. Res.* **84**, 5535–9.

Brett, R. (1976). The current status of speculations on the composition of the core of the Earth. *Rev. Geophys. Space Phys.* **14**, 375–83.

Brillouin, L. (1938). On the thermal dependence of elasticity in solids. *Phys. Rev.* **54**, 916–17.

(1940). Influence de la température sur l'élasticité d'un solide. *Mémorial des Sciences Mathématiques* **94**, 1–65.

(1960). *Les Tenseurs en Mécanique et en Elasticité*. Paris: Masson.

Brillouin, M. (1898). Théorie de la fusion complète et de la fusion pâteuse. *Ann. Chim. et Phys.* **13**, 264–75.

Brown, J. M. (1986). Interpretation of the D'' zone at the base of the mantle: dependence on assumed values of thermal conductivity. *Geophys. Res. Lett.* **13**, 1509–12.

Brown, J. M., and McQueen, R. G. (1980). Melting of iron under core conditions. *Geophys. Res. Lett.* **7**, 533–6.

(1982). The equation of state for iron and the Earth's core. In *High pressure research in geophysics,* S. Akimoto and M. H. Manghnani, eds. Dordrecht: Reidel, 611–23.

(1986). Phase transitions, Grüneisen parameter and elasticity for shocked iron between 77 GPa and 400 GPa. *J. Geophys. Res.* **91**, 7485–94.

Brown, J. M., and Shankland, T. J. (1981). Thermodynamic parameters in the Earth as determined from seismic profiles. *Geophys. J. R. Astr. Soc.* **66**, 579–96.

Brush, S. G. (1962). Theories of liquid viscosity. *Chemical Reviews* **62**, 513–48.

(1979). Nineteenth century debates about the inside of the Earth: Solid, liquid or gas? *Annals of Science* **36**, 225–54.

(1982). Chemical history of the Earth's core. *EOS Trans AGU* **63**, 1185–8.

Bukowinski, M. S. T. (1977). A theoretical equation of state for the inner core. *Phys. Earth Planet. Interiors* **14**, 333–44.

Bullen, K. E. (1963). An index of degree of chemical inhomogeneity in the Earth. *Geophys. J. R. Astr. Soc.* **7**, 584–92.

(1975). *The Earth's density*. London: Chapman and Hall.

Bullen, K. E., and Bolt, B. A. (1985). *An introduction to the theory of seismology*. Cambridge: Cambridge University Press.

Bundy, F. P. (1965). Pressure–temperature phase diagram of iron to 200 kbar, 900°C. *J. Appl. Phys.* **36**, 616–20.

Burnley, P. C., and Green, H. W. (1989). The olivine–spinel transformation: The dependence of the nucleation mechanism on the level of stress. *Nature* **338**, 753–6.

Butler, R., and Anderson, D. L. (1978). Equation of state fits to the lower mantle and outer core. *Phys. Earth Planet. Interiors* **17**, 147–62.

Callen, H. B. (1985). *Thermodynamics, an introduction to thermostatistics*. New York: Wiley.

Cannon, W. R., and Langdon, T. G. (1983). Creep of ceramics. *J. Materials Sci.* **18**, 1–50.

Carslaw, H. S., and Jaeger, J. C. (1959). *Conduction of heat in solids*. Oxford: Oxford University Press.

Cazenave, A.; Souriau, A.; and Dominh, K. (1989). Global coupling of Earth surface topography with hotspots, geoid and mantle heterogeneities. *Nature* **340**, 54–7.

Cemic, L.; Will, G.; and Hinze, E. (1980). Electrical conductivity measurements on olivines Mg_2SiO_4–Fe_2SiO_4 under defined thermodynamic conditions. *Phys. Chem. Minerals* **6**, 95–107.

Chen, H. C.; Gartstein, E.; and Mason, T. O. (1982). Conduction mechanism analysis for $Fe_{1-\delta}O$ and $Co_{1-\delta}O$. *J. Phys. Chem. Solids* **43**, 991–5.

Chen, W. K., and Peterson, N. L. (1980). Iron diffusion and electrical conductivity in magnesiowüstite solid solutions $(Mg,Fe)O$. *J. Phys. Chem. Solids* **41**, 335–9.

Chopelas, A., and Boehler, R. (1989). Thermal expansion measurements at very high pressures, systematics and a case for a chemically homogeneous mantle. *Geophys. Res. Lett.* **16**, 1347–50.

Christensen, U. (1984). Large-scale dynamics and the 670 km discontinuity. *Terra Cognita* **4**, 59–66.

Chung, D. H. (1972). Birch's law: Why is it so good? *Science* **177**, 261–3.

Clark, S. P. (1957). Radiative transfer in the Earth's mantle. *Trans. Amer. Geophys. Union* **38**, 931–8.

Cohen, R. E. (1987). Elasticity and equation of state of $MgSiO_3$ perovskite. *Geophys. Res. Lett.* **14**, 1053–6.

Cohen, M. H., and Turnbull, D. S. (1959). Molecular transport in liquids and glasses. *J. Chem. Phys.* **31**, 1164–9.

Cotterill, R. M. J. (1980). The physics of melting. *J. Crystal Growth* **48**, 582–8.

Coey, J. M. D.; Bakas, T.; McDonagh, C. M.; and Litterst, F. J. (1989). Electrical and magnetic properties of cronstedtite. *Phys. Chem. Minerals* **16**, 394–400.

Davies, G. F. (1976). The estimation of elastic properties from analogue compounds. *Geophys. J. R. Astr. Soc.* **44**, 625–47.

Davies, G. F., and Gurnis, M. (1986). Interaction of mantle dregs with convection: lateral heterogeneity at the core–mantle boundary. *Geophys. Res. Lett.* **13**, 1517–20.

Dennis, P. F. (1984). Oxygen self-diffusion in quartz under hydrothermal conditions. *J. Geophys. Res.* **89**, 4047–57.

Dieckmann, R. (1984). Point defects and transport properties of binary and ternary oxides. In *Transport in non-stoichiometric compounds,* G. Petot-Ervas, H. J. Matzke, and C. Monty, eds. Amsterdam: North-Holland, 1–22.

Domb, C. (1951). The melting curve at high pressures. *Philos. Mag.* **42**, 1316–24.

Dornboos, D. J.; Spiliopoulos, S.; and Stacey, F. D. (1986). Seismological properties of D'' and the structure of a thermal boundary layer. *Phys. Earth Planet. Interiors* **41**, 225–39.

Dosdale, T., and Brook, R. J. (1983). Comparison of diffusion data and of activation energies. *J. Amer. Ceram. Soc.* **66**, 392–5.

Drickamer, H. G. (1963). The electronic structure of solids under pressure. In *Solids under pressure,* W. Paul and D. Warschauer, eds. New York: McGraw-Hill, 357–84.

Duba, A., and Nicholls, I. A. (1973). The influence of the oxidation state on the electrical conductivity of olivine. *Earth Planet. Sci. Lett.* **18**, 59–64.

Duba, A.; Ito, J.; and Jamieson, J. C. (1973). The effect of ferric iron on the electrical conductivity of olivine. *Earth Planet. Sci. Lett.* **18**, 279–84.

Duba, A.; Heard, H. C.; and Schock, R. N. (1974). Electrical conductivity of olivine at high pressure and under controlled oxygen fugacity. *J. Geophys. Res.* **79**, 1667–73.

Duffy, T. S., and Anderson, D. L. (1989). Seismic velocities in mantle minerals and the mineralogy of the upper mantle. *J. Geophys. Res.* **94**, 1895–912.

Dugdale, J. S., and MacDonald, D. K. C. (1953). The thermal expansion of solids. *Phys. Rev.* **89**, 832–4.

(1955). Lattice thermal conductivity. *Phys. Rev.* **98**, 1751–2.

Durand, M. A. (1936). The temperature variation of the elastic moduli of NaCl, KCl, and MgO. *Phys. Rev.* **50**, 449–55.

Dziewonski, A. M., and Anderson, D. L. (1981). Preliminary reference Earth model. *Phys. Earth Planet. Interiors* **25**, 297–356.

Eliezer, S.; Ghatak, A.; and Hora, H. (1986). *An introduction to equations of state. Theory and applications.* Cambridge: Cambridge University Press.

Ellsworth, K.; Schubert, G.; and Sammis, C. G. (1985). Viscosity profile of the lower mantle. *Geophys. J. R. Astr. Soc.* **83**, 199–214.

Elsasser, W. M. (1951). Quantum-theoretical densities of solids at extreme compressions. *Science,* **113**, 105–7.

Emin, D. (1975). Transport properties of small polarons. *J. Sol. State Chem.* **12**, 246–52.

Exner, O. (1964). Concerning the isokinetic relation. *Nature* **201**, 488–90.

Eyring, H. (1936). Viscosity, plasticity and diffusion as examples of absolute reaction rates. *J. Chem. Phys.* **4**, 283–91.

Falzone, A. J., and Stacey, F. D. (1980). Second-order elasticity theory: Explanation for the high Poisson's ratio of the inner core. *Phys. Earth Planet. Interiors* **21**, 371–7.

Feynman, R. P.; Metropolis, N.; and Teller, E. (1949). Equations of state of elements based on the generalized Fermi–Thomas theory. *Phys. Rev.* **75**, 1561–73.

Fischer, K. M.; Jordan, T. H.; and Creager, K. C. (1988). Seismic constraints on the morphology of deep slabs. *J. Geophys. Res.* **93**, 4773–83.

Frank, F. C. (1939). Melting as a disorder phenomenon. *Proc. Roy. Soc.* **170**, 182–9.

Freer, R. (1980). Self-diffusion and impurity diffusion in oxides. *J. Materials Sci.* **15**, 803–24.

(1981). Diffusion in silicate minerals and glasses: A data digest and guide to the literature. *Contrib. Mineral. Petrol.* **76**, 440–54.

Friedel, J. (1964). *Dislocations.* Oxford: Pergamon Press.

Frost, H. J., and Ashby, M. F. (1982). *Deformation-mechanism maps.* Oxford: Pergamon Press.

Fujisawa, H.; Fujii, N.; Mizutani, H.; Kanamori, H.; and Akimoto, S. (1968). Thermal diffusivity of Mg_2SiO_4, Fe_2SiO_4 and NaCl at high pressures and temperatures. *J. Geophys. Res.* **73**, 4727–33.

Fukai, Y., and Suzuki, T. (1986). Iron–water reaction under high pressure and its implications in the evolution of the Earth. *J. Geophys. Res.* **91**, 9222–30.

Furnish, M. D., and Bassett, W. A. (1983). Investigation of the mechanism of the olivine–spinel transition in fayalite by synchrotron radiation. *J. Geophys. Res.* **88**, 10333–41.

Giletti, B. J., and Hess, K. C. (1988). Oxygen diffusion in magnetite. *Earth Planet Sci. Lett.* **89**, 115–22.

Giletti, B. J., and Yund, R. A. (1984). Oxygen diffusion in quartz. *J. Geophys. Res.* **89**, 4039–46.

Gillet, P.; Guyot, F.; and Malezieux, J. M. (1989). High pressure, high temperature Raman spectroscopy of Ca_2GeO_4 (olivine form): Some insights on anharmonicity. *Phys. Earth Planet. Interiors* **58**, 141–54.

Gilvarry, J. J. (1956a). The Lindemann and Grüneisen laws. *Phys. Rev.* **102**, 308–16.

———(1956b). Grüneisen's law and the fusion curve at high pressure. *Phys. Rev.* **102**, 317–25.

———(1956c). Equation of the fusion curve. *Phys. Rev.* **102**, 325–31.

———(1956d). Grüneisen parameter for a solid under finite strain. *Phys. Rev.* **102**, 331–40.

———(1957a). Temperature dependent equations of state of solids. *J. Appl. Phys.* **28**, 1253–61.

———(1957b). Temperatures in the Earth's interior. *J. Atm. Terrestr. Phys.* **10**, 84–95.

———(1966). Lindemann and Grüneisen laws and a melting law at high temperatures. *Phys. Rev. Lett.* **16**, 1089–91.

Graham, E. K., and Barsch, G. R. (1969). Elastic constants of single-crystal forsterite as a function of temperature and pressure. *J. Geophys. Res.* **74**, 5949–60.

Graham, E. K.; Schwab, J. A.; Sopkin, S. M.; and Takei, H. (1988). The pressure and temperature dependence of the elastic properties of single-crystal fayalite Fe_2SiO_4. *Phys. Chem. Minerals* **16**, 186–98.

Grosse, A. V. (1963). The empirical relation between the activation energy of viscosity of liquid metals and their melting points. *J. Inorg. Nuclear Chem.* **25**, 317–18.

Guillermet, A. F., and Gustafson, P. (1984). An assessment of the thermodynamic properties and phase diagram of iron. *Materials Center, Royal Inst. Technology, Stockholm, Report* Trita-Mac-0229.

Gurney, R. W. (1966). *Introduction to statistical mechanics.* New York: Dover.

Guyot, F.; Madon, M.; Peyronneau, J.; and Poirier, J. P. (1988). X-ray microanalysis of high-pressure/high-temperature phases synthesized from natural olivine in a diamond anvil cell. *Earth Planet. Sci. Lett.* **90**, 52–64.

Guyot, F.; Peyronneau, J.; and Poirier, J. P. (1988). TEM study of high pressure reactions between iron and silicate perovskites. *Chemical Geology* **70**, 61.

Hager, B. H.; Clayton, R. W.; Richards, M. A.; Comer, R. P.; and Dziewonski, A. (1985). Lower mantle heterogeneity, dynamic topography and the geoid. *Nature* **313**, 541–5.

Hamilton, R. M. (1965). Temperature variation at constant pressures of the electrical conductivity of periclase and olivine. *J. Geophys. Res.* **70**, 5679–92.

Hansen, K. W., and Cutler, I. B. (1966). Electrical conductivity in $Fe_{1-x}O$–MgO solid solutions. *J. Amer. Ceram. Soc.* **49**, 100–2.

Harris, P. S. (1973). Compensation effect and experimental error. *Nature* **243**, 401–2.

Hart, S. R. (1981). Diffusion compensation in natural silicates. *Geochim. Cosmochim. Acta* **45**, 279–91.

Hart, S. R., and Zindler, A. (1986). In search of a bulk Earth composition. *Chemical Geology* **57**, 247–67.

Heinz, D. L., and Jeanloz, R. (1983). Inhomogeneity parameter of a homogeneous Earth. *Nature* **301**, 138–9.

(1987). Measurement of the melting curve of $Mg_{0.9}Fe_{0.1}SiO_3$ at lower mantle conditions and its geophysical implications. *J. Geophys. Res.* **92**, 11437–44.

Hemley, R. J.; Bell, P. M.; and Mao, H. K. (1987). Laser techniques in high-pressure geophysics. *Science* **237**, 605–12.

Hemley, R. J.; Jackson, M. D.; and Gordon, R. G. (1985). First-principles theory for the equations of state of minerals at high pressures and temperatures: application to MgO. *Geophys. Res. Lett.* **12**, 247–50.

(1987). Theoretical study of the structure, lattice dynamics, and equation of state of perovskite-type $MgSiO_3$ and $CaSiO_3$. *Phys. Chem. Minerals* **14**, 2–12.

Hemley, R. J.; Jephcoat, A. P.; Mao, H. K.; Ming, L. C.; and Manghnani, M. H. (1988). Pressure induced amorphization of silica. *Nature* **334**, 52–4.

Higgins, G., and Kennedy, G. C. (1971). The adiabatic gradient and the melting point gradient in the core of the Earth. *J. Geophys. Res.* **76**, 1870–8.

Honig, J. M. (1970). Band and transport theories in solids. In *Modern aspects of solid state chemistry,* C. N. R. Rao, ed. New York: Plenum, 497–566.

Horai, K. (1970). Thermal conductivity of rock-forming minerals. *J. Geophys. Res.* **76**, 1278–1308.

Horai, K., and Shankland, T. (1987). Thermal conductivity of rocks and minerals. In *Methods of experimental physics, geophysics,* C. G. Sammis and T. L. Henyey, eds. New York: Academic, **24 A**, 271–302.

Horai, K., and Simmons, G. (1969). Thermal conductivity of rock-forming minerals. *Earth Planet. Sci. Lett.* **6**, 359–68.

(1970). An empirical relationship between thermal conductivity and Debye temperature for silicates. *J. Geophys. Res.* **75**, 978–82.

Irifune, T. (1987). An experimental investigation of the pyroxene garnet transformation in a pyrolite composition and its bearing on the composition of the mantle. *Phys. Earth Planet. Interiors* **45**, 324–36.

Irvine, R. D., and Stacey, F. D. (1975). Pressure dependence of the thermal Grüneisen parameter, with application to the lower mantle and outer core. *Phys. Earth Planet. Interiors* **11**, 157–65.

Ito, E., and Katsura, T. (1989). A temperature profile of the mantle transition zone. *Geophys. Res. Lett.* **16**, 425–8.

Ito, E., and Matsui, Y. (1978). Synthesis and crystal-chemical characterization of $MgSiO_3$ perovskite. *Earth Planet. Sci. Lett.* **38**, 443–50.

Ito, E., and Navrotsky, A. (1985). MgSiO ilmenite: calorimetry, phase equilibria, and decomposition at atmospheric pressure. *Amer. Mineral.* **70**, 1020–6.

Ito, E., and Takahashi, E. (1987a). Ultra-high pressure phase transformations and the constitution of the deep mantle. In *High-pressure research in mineral physics,* M. H. Manghnani and Y. Syono, eds. Washington, D.C.: American Geophysical Union, 221–9.

(1987b). Melting of peridotite at uppermost lower mantle conditions. *Nature* **328**, 514–17.

(1989). Post-spinel transformations in the system Mg_2SiO_4–Fe_2SiO_4 and some geophysical implications. *J. Geophys. Res.* **94**, 10637–46.

Ito, E., and Yamada, H. (1982). Stability relations of silicate spinels, ilmenites and perovskites. In *High-pressure research in geophysics,* S. Akimoto and M. H. Manghnani, eds. Dordrecht: Reidel, 405–19.

Ito, E.; Takahashi, E.; and Matsui, Y. (1984). The mineralogy and chemistry of the lower mantle: An implication of the ultra-high pressure phase relations in the system $MgO-FeO-SiO_2$. *Earth Planet. Sci. Lett.* **67**, 238–48.

Iyengar, G. N. K., and Alcock, C. B. (1970). A study of semiconduction in dilute magnesiowüstites. *Philos. Mag.* **21**, 293–304.

Jackson, I. (1977). Melting of some alkaline–earth and transition–metal fluorides and alkali fluoberyllates at elevated pressures: a search for melting systematics. *Phys. Earth Planet. Interiors* **14**, 143–64.

(1983). Some geophysical constraints on the chemical constitution of the Earth's lower mantle. *Earth Planet. Sci. Lett.* **62**, 91–103.

Jackson, I. N. S., and Liebermann, R. C. (1974). Melting and elastic shear instability of alkali halides. *J. Phys. Chem. Solids* **35**, 115–19.

Jackson, I., and Niesler, H. (1982). The elasticity of periclase to 3GPa and some geophysical implications. *Advances in Earth and Planet. Sci.* **12**, 93–113.

Jacobs, J. A. (1986). *The Earth's core.* New York: Academic.

Jaoul, O.; Houlier, B.; and Abel, F. (1983). Study of ^{18}O diffusion in magnesium orthosilicate by nuclear microanalysis. *J. Geophys. Res.* **88**, 613–24.

Jaoul, O.; Poumellec, M.; Froidevaux, C.; and Havette, A. (1981). Silicon diffusion in forsterite: A new constraint for understanding mantle deformation. In *Anelasticity in the Earth,* F. D. Stacey, M. S. Paterson, and A. Nicolas, eds. Washington, D.C.: American Geophysical Union, 95–100.

Jeanloz, R. (1981). Majorite: Vibrational and compressional properties of a high-pressure phase. *J. Geophys. Res.* **86**, 6171–9.

Jeanloz, R., and Grover, R. (1988). Birch–Murnaghan and U_s-u_p equations of state. In *Shock waves in condensed matter 1987,* S. C. Schmitt and N. C. Holmes, eds. Amsterdam: Elsevier.

Jeanloz, R., and Morris, S. (1986). Temperature distribution in the crust and mantle. *Ann. Rev. Earth Planet. Sci.* **14**, 377–415.

Jeanloz, R., and Richter, F. M. (1979). Convection, composition and the thermal state of the lower mantle. *J. Geophys. Res.* **84**, 5497–503.

Jeanloz, R., and Thompson, A. B. (1983). Phase transitions and mantle discontinuities. *Rev. Geophys. Space Phys.* **21**, 51–74.

Jephcoat, A., and Olson, P. (1987). Is the inner core of the Earth pure iron? *Nature* **325**, 332–5.

Jones, L. E. A., and Liebermann, R. C. (1974). Elastic and thermal properties of fluoride and oxide analogues in the rocksalt, fluorite, rutile and perovskite structures. *Phys. Earth Planet. Interiors* **9**, 101–7.

Kanamori, H.; Fujii, N.; and Mizutani, H. (1968). Thermal diffusivity measurement of rock-forming minerals from 300 to 1100 K. *J. Geophys. Res.* **73**, 595–605.

Kapusta, B., and Guillopé, M. (1988). High ionic diffusivity in the perovskite $MgSiO_3$: A molecular dynamics study. *Philos. Mag.* **58**, 809–16.

Karato, S. (1981a). Rheology of the lower mantle. *Phys. Earth Planet. Interiors* **24**, 1–14.

(1981b). Pressure dependence of diffusion in ionic solids. *Phys. Earth Planet. Interiors* **25**, 38–51.

Kato, T., and Kumazawa, M. (1985a). Garnet phase of $MgSiO_3$ filling the pyroxene ilmenite gap at very high temperature. *Nature* **316**, 803–5.

(1985b). Effect of high pressure on the melting relation in the system Mg₂SiO₄–MgSiO₃. *J. Phys. Earth* **33**, 513–24.

Katsura, T., and Ito, E. (1989). The system Mg₂SiO₄–Fe₂SiO₄ at high pressures and temperatures: precise determination of stabilities of olivine, modified spinel and spinel. *J. Geophys. Res.* **94**, 15663–70.

Kawai, N., and Inokuti, Y. (1968). Low temperature melting of elements under high pressure and its progression in the periodic table. *Jap. J. Appl. Phys.* **7**, 989–1004.

Kemeny, G., and Rosenberg, B. (1973). Compensation law in thermodynamics and thermal death. *Nature* **243**, 400.

Kennedy, G. C., and Vaidya, S. N. (1970). The effect of pressure on the melting temperature of solids. *J. Geophys. Res.* **75**, 1019–22.

Kenyon, P. M., and Turcotte, D. L. (1983). Convection in a two-layer mantle with a strongly temperature-dependent viscosity. *J. Geophys. Res.* **88**, 6403–14.

Keyes, R. W. (1958). Volumes of activation for diffusion in solids. *J. Chem. Phys.* **29**, 467–75.

(1960). Volumes of activation. II. Pressure dependence of activation parameters. *J. Chem. Phys.* **32**, 1066–7.

(1963). Continuum model of the effect of pressure on activated processes. In *Solids under pressure,* W. Paul and D. Warschauer, eds. New York: McGraw-Hill, 71–91.

Kieffer, S. W. (1976). Lattice thermal conductivity within the Earth and considerations of a relationship between the pressure dependence of the thermal diffusivity and the volume dependence of the Grüneisen parameter. *J. Geophys. Res.* **81**, 3025–30.

(1979). Thermodynamics and lattice vibrations of minerals:
    a) 1. Mineral heat capacities and their relationships to simple lattice vibrational models. *Rev. Geophys. Space Phys.* **17**, 1–19.
    b) 2. Vibrational characteristics of silicates. *Ibid.* **17**, 20–34.
    c) 3. Lattice dynamics and an approximation for minerals with application to simple substances and framework silicates. *Ibid.* **17**, 35–59.

(1985). Heat capacity and entropy: systematic relations to lattice vibrations. In *Microscopic to macroscopic,* S. W. Kieffer and A. Navrotsky, eds. Washington, D.C.: Mineral. Soc. America, 65–126.

Kirchheim, R., and Huang, X. Y. (1987). A relationship between prefactor and activation energy for diffusion. *Phys. stat. sol. (b)* **144**, 253–7.

Kittel, C. (1967). *Introduction to solid state physics.* New York: Wiley.

Knittle, E., and Jeanloz, R. (1986). High-pressure metallization of FeO and implications for the Earth's core. *Geophys. Res. Lett.* **13**, 1541–4.

(1987). Synthesis and equation of state of (Mg,Fe)SiO₃ perovskite to over 100 gigapascals. *Science* **235**, 668–70.

(1989). Simulating the core–mantle boundary: An experimental study of high-pressure reactions between silicate and liquid iron. *Geophys. Res. Lett.* **16**, 609–12.

Knittle, E.; Jeanloz, R.; and Smith, G. L. (1987). Thermal expansion of silicate perovskite and stratification of the Earth's mantle. *Nature* **319**, 214–16.

Knopoff, L., and Shapiro, J. N. (1969). Comments on the interrelationships between Grüneisen's parameter and shock and isothermal equations of state. *J. Geophys. Res.* **74**, 1439–50.

Kobayashi, Y., and Maruyama, H. (1971). Electrical conductivity of olivine single crystals at high temperature. *Earth Planet. Sci. Lett.* **11**, 415–19.

Kofstad, P. (1983). *Nonstoichiometry, diffusion and electrical conductivity in binary metal oxides.* Malabar, FL: Krieger.

Kraut, E. A., and Kennedy, G. C. (1966a). New melting law at high pressures. *Phys. Rev. Lett.* **16**, 608–9.

(1966b). New melting law at high pressures. *Phys. Rev.* **151**, 668–75.

Kudoh, Y.; Ito, E.; and Takeda, H. (1987). Effect of pressure on the crystal structure of perovskite-type $MgSiO_3$. *Phys. Chem. Minerals* **14**, 350–4.

Kündig, W., and Hargrove, R. S. (1969). Electron hopping in magnetite. *Solid State Comm.* **7**, 223–7.

Kuhlman-Wilsdorf, D. (1965). Theory of melting. *Phys. Rev.* **A 140**, 1599–1610.

Kumazawa, M., and Anderson, O. L. (1969). Elastic moduli, pressure derivatives, and temperature derivatives of single-crystal olivine and single-crystal forsterite. *J. Geophys. Res.* **74**, 5961–72.

Lacam, A.; Madon, M.; and Poirier, J. P. (1980). Olivine glass and spinel formed in a laser-heated, diamond–anvil high pressure cell. *Nature* **288**, 155–7.

Laplace, M. le Marquis de (1825). *Traité de Mécanique céleste,* t. v, 48–50, Paris: Bachelier.

Lasaga, A. C. (1979). Multicomponent exchange and diffusion in silicates. *Geochim. Cosmochim. Acta* **43**, 455–69.

Lasocka, M. (1975). On the entropy of melting. *Physics Lett.* **51A**, 137–8.

Lauterjung, J., and Will, G. (1986). The kinetics of the olivine–spinel transformation in Mg_2GeO_4 under high pressure and temperature. *Physica* **139–140B**, 343–6.

Lawson, A. W. (1957). On the high temperature heat conductivity of insulators. *Phys. and Chem. Solids* **3**, 155.

Lay, T. (1989). Structure of the core–mantle transition zone: A chemical and thermal boundary layer. *EOS Trans. AGU* **70**, 49–59.

Lees, A. C.; Bukowinski, M. S. T.; and Jeanloz, R. (1983). Reflection properties of phase transition and compositional change models of the 670 km discontinuity. *J. Geophys. Res.* **88**, 8145–59.

Lennard-Jones, J. E., and Devonshire, A. F. (1937). Critical phenomena in gases–I. *Proc. Roy. Soc.* **A 163**, 53–70.

(1939). Critical and cooperative phenomena.

    a) III. A theory of melting and the structure of liquids. *Proc. Roy. Soc.* **A 169**, 317–38.

    b) IV. A theory of disorder in solids and liquids and the process of melting. *Proc. Roy. Soc.* **A 170**, 464–84.

Levin, E. M.; Robbins, C. R.; and McMurdie, H. F. (1964). *Phase diagrams for ceramists.* Columbus, OH: American Ceramic Society.

Li, X., and Jeanloz, R. (1987). Measurement of the electrical conductivity of $(Mg,Fe)SiO_3$ perovskite and a perovskite-dominated assemblage at lower mantle conditions. *Geophys. Res. Lett.* **14**, 1075–8.

Libby, W. F. (1966). Melting points at high compressions from zero compression properties through the Kennedy relation. *Phys. Rev. Lett.* **17**, 423–4.

Liebermann, R. C. (1973). On velocity–density systematics, polymorphic phase transformations and the transition zone of the Earth's mantle. *Comments on Earth Sciences: Geophysics* **3**, 127–33.

(1982). Elasticity of minerals at high pressure and temperature. In *High pressure researches in geosciences,* W. Schreyer, ed. Stuttgart: Schweizerbart'sche, 1–14.

Liebermann, R. C., and Ringwood, A. E. (1973). Birch's law and polymorphic phase transformations. *J. Geophys. Res.* **78**, 6926–32.

Liebermann, R. C.; Jones, L. E. A.; and Ringwood, A. E. (1977). Elasticity of aluminate, titanate, stannate and germanate compounds with the perovskite structure. *Phys. Earth Planet. Interiors* **14**, 165–78.

Liempt, J. A. M. van (1935). Die Berechnung der Auflockerungswärme der Metalle aus Rekristallisationsdaten. *Z. Phys.* **96**, 534–41.

Lindemann, F. A. (1910). Ueber die Berechnung molekularer Eigenfrequenzen. *Physikalisches Zeitsch.* **11**, 609–12.

Liu, L. G. (1974). Silicate perovskite from phase transformations of pyrope garnet at high pressure and temperature. *Geophys. Res. Lett.* **1**, 277–80.

(1975a). Post-oxide phases of forsterite and enstatite. *Geophys. Res. Lett.* **2**, 417–19.

(1975b). On the (γ, ϵ, l) triple point of iron and the Earth's core. *Geophys. J. R. Astr. Soc.* **43**, 697–705.

(1976). The high-pressure phases of $MgSiO_3$. *Earth Planet. Sci. Lett.* **31**, 200–8.

(1979). On the 650 km seismic discontinuity. *Earth Planet. Sci. Lett.* **42**, 202–8.

(1982). Chemical inhomogeneity of the mantle: Geochemical considerations. *Geophys. Res. Lett.* **9**, 124–6.

(1987). New silicate perovskites. *Geophys. Res. Lett.* **14**, 1077–82.

Liu, L. G., and Bassett, W. A. (1975). The melting of iron up to 200 kbar. *J. Geophys. Res.* **80**, 3777–82.

(1986). *Elements, oxides, silicates.* Oxford: Oxford University Press.

Liu, L. G., and Ringwood, A. E. (1975). Synthesis of a perovskite-type polymorph of $CaSiO_3$. *Earth Planet. Sci. Lett.* **28**, 209–11.

Loper, D. E. (1984). The dynamical structure of D'' and deep plumes in a non-Newtonian mantle. *Phys. Earth Planet. Interiors* **34**, 57–67.

Love, A. E. H. (1944). *A treatise on the mathematical theory of elasticity.* New York: Dover.

MacKenzie, D. P. (1967). The viscosity of the mantle. *Geophys. J. R. Astr. Soc.* **14**, 297–305.

McQueen, R. G.; Fritz, J. N.; and Marsh, S. P. (1963). On the equation of state of stishovite. *J. Geophys. Res.* **68**, 2319–22.

(1964). On the composition of the Earth's interior. *J. Geophys. Res.* **69**, 2947–65.

Madon, M., and Poirier, J. P. (1983). Transmission electron microscope observations of α, β and γ $(MgFe)_2SiO_4$ in shocked meteorites: planar defects and polymorphic transitions. *Phys. Earth Planet. Interiors* **33**, 31–44.

Madon, M.; Castex, J.; and Peyronneau, J. (1989). A new hollandite-type structure as a possible host for calcium and aluminium in the lower mantle. *Nature* **342**, 422–4.

Madon, M.; Guyot, F.; Peyronneau, J.; and Poirier, J. P. (1989). Electron microscopy of high-pressure phases synthesized from natural olivine in diamond anvil cell. *Phys. Chem. Minerals* **16**, 320–30.

Maj, S. (1978). A relationship between phonon conductivity and seismic parameter for silicate minerals. *Pageoph* **116**, 1225–30.

Mao, H. K. (1972). Observations of optical absorption and electrical conductivity in magnesiowüstite at high pressures. *Carnegie Institution of Washington Annual Report,* 554–7.

(1976). Charge transfer processes at high pressure. In *The physics and chemistry of minerals and rocks,* R. G. J. Strens, ed. London: Wiley, 573–81.

Mao, H. K., and Bell, P. M. (1979). Equations of state of MgO and ϵ-Fe under static pressure conditions. *J. Geophys. Res.* **84**, 4533–6.

Mao, H. K.; Bell, P. M.; and Hadidiacos, C. (1987). Experimental phase relations of iron to 360 kbar, 1400°C, determined in an internally heated diamond–anvil apparatus. In *High-pressure research in mineral physics,* M. H. Manghnani and Y. Syono, eds. Washington, D.C.: American Geophysical Union, 135–8.

Martin, C. J., and O'Connor, D. A. (1977). An experimental test of Lindemann law. *J. Phys. C* **10**, 3521–6.

Mashimo, T.; Kondo, K.; Sawaoka, A.; Syono, Y.; Takei, H.; and Ahrens, T. J. (1980). Electrical conductivity measurement of fayalite under shock compression up to 56 GPa. *J. Geophys. Res.* **85**, 1876–81.

Masters, G. (1979). Observational constraints on the chemical and thermal structure of the Earth's deep interior. *Geophys. J. R. Astr. Soc.* **57**, 507–34.

May, A. N. (1970). Extrapolation of the shear elastic moduli of face-centred cubic solids to the molten state. *Nature* **228**, 990–1.

Means, W. D. (1976). *Stress and strain.* New York: Springer.

Mercier, J. C. C. (1980). Single pyroxene thermobarometry. *Tectonophysics* **70**, 1–37.

Ming, L. C., and Bassett, W. A. (1974). Laser heating in the diamond anvil press up to 2000°C sustained and 3000°C pulsed at pressures up to 260 kilobars. *Rev. Sci. Instr.* **45**, 1115–18.

(1975). The post-spinel phases in the Mg_2SiO_4–Fe_2SiO_4 system. *Science* **187**, 66–8.

Mishima, O.; Calvert, L. D.; and Whalley, E. (1984). "Melting ice" at 77K and 10 kbar: a new method of making amorphous solids. *Nature* **310**, 393–5.

Mitoff, S. P. (1962). Electronic and ionic conductivity in single crystals of MgO. *J. Chem. Phys.* **36**, 1383–9.

Mizushima, S. (1960). Dislocation models of liquid structure. *J. Phys. Soc. Japan* **15**, 70–7.

Mizutani, H., and Kanamori, H. (1967). Electrical conductivities of rock forming minerals at high temperatures. *J. Phys. Earth* **15**, 25–31.

Morelli, A., and Dziewonski, A. M. (1987). Topography of the core–mantle boundary and lateral homogeneity of the liquid core. *Nature* **325**, 678–83.

Morin, F. J.; Oliver, J. R.; and Housley, R. M. (1977). Electrical properties of forsterite, Mg_2SiO_4, I. *Phys. Rev.* **B16**, 4434–45.

(1979). Electrical properties of forsterite, Mg_2SiO_4, II. *Phys. Rev.* **B19**, 2886–94.

Mott, N. F. (1961). The transition to the metallic state. *Philos. Mag.* **6**, 287–309.

Mott, N. F., and Jones, H. (1958). *The theory of the properties of metals and alloys.* New York: Dover.

Muirhead, K. (1985). Comments on "Reflection properties of phase transition and compositional change models of the 670 km discontinuity" by Lees, Bukowinski and Jeanloz. *J. Geophys. Res.* **90**, 2057–9.

Mukherjee, K. (1966). Clapeyron's equation and melting under high pressures. *Phys. Rev. Lett.* **17**, 1252–4.

Mulargia, F. (1986). The physics of melting and the temperatures in the Earth's outer core. *Q. Jl. R. Astr. Soc.* **27**, 383–402.

Mulargia, F., and Quareni, F. (1988). Validity of the Sutherland–Lindemann law and melting temperatures in the Earth's interior. *Geophys. J.* **92**, 269–82.

Muller, O., and Roy, R. (1974). *The major ternary structural families.* Berlin: Springer.

Murnaghan, F. D. (1967). *Finite deformation of an elastic solid.* New York: Dover.

Nabarro, F. R. N. (1967). *Theory of crystal dislocations.* Oxford: Oxford University Press.

Nachtrieb, N. H. (1967). Self-diffusion in liquid metals. *Adv. in Phys.* **16**, 309–23.
  (1977). Atomic transport properties in liquid metals. In *The properties of liquid metals,* S. Takeuchi, ed. London: Taylor and Francis.

Navrotsky, A. (1980). Lower mantle phase transitions may generally have negative pressure–temperature slopes. *Geophys. Res. Lett.* **7**, 709–11.

Ninomiya, T. (1978). Theory of melting, dislocation model. *J. Phys. Soc. Japan* **44**, 263–71.

Nye, J. F. (1957). *Physical properties of crystals.* Oxford: Oxford University Press.

O'Connell, R. J. (1977). On the scale of mantle convection. *Tectonophysics* **38**, 119–36.

Oishi, Y., and Ando, K. (1984). Oxygen self-diffusion coefficient in single crystal forsterite. In *Materials science of the Earth's interior,* I. Sunagawa, ed. Tokyo: Terrapub, 271–80.

O'Keefe, M., and Bovin, J. O. (1979). Solid electrolyte behavior of $NaMgF_3$: geophysical implications. *Science* **206**, 599–600.

O'Keefe, M., and Hyde, B. G. (1981). Why olivine transforms to spinel at high pressure. *Nature* **293**, 727–8.

Olson, P.; Schubert, G.; and Anderson, C. (1987). Plume formation in the D'' layer and the roughness of the core–mantle boundary. *Nature* **327**, 409–13.

Olson, P.; Silver, P. G.; and Carlson, R. W. (1990). The large-scale structure of convection in the Earth's mantle. *Nature* **344**, 209–15.

Oriani, R. A. (1951). The entropies of melting of metals. *J. Chem. Phys.* **19**, 93–7.

Parkhomenko, E. I. (1982). Electrical resistivity of minerals and rocks at high temperature and pressure. *Rev. Geophys. Space Phys.* **20**, 193–218.

Paul, W., and Warschauer, D. M. (1963). The role of pressure in semiconductor research. In *Solids under pressure,* W. Paul and D. Warschauer, eds. New York: McGraw-Hill, 179–249.

Peltier, W. R., and Jarvis, G. T. (1982). Whole mantle convection and the thermal evolution of the Earth. *Phys. Earth Planet. Interiors* **29**, 281–304.

Peyronneau, J., and Poirier, J. P. (1989). Electrical conductivity of the material of the Earth's lower mantle. *Nature* **342**, 537-9.

Poirier, J. P. (1978). Is power-law creep diffusion controlled? *Acta Metall.* **26**, 629-37.

(1981a). Martensitic olivie–spinel transformation and plasticity of the mantle transition zone. *Anelastic properties and related processes in the Earth's mantle,* American Geophysical Union monograph.

(1981b). On the kinetics of the olivine spinel transition. *Phys. Earth Planet. Interiors* **26**, 179-87.

(1985). *Creep of crystals.* Cambridge: Cambridge University Press.

(1986). Dislocation-mediated melting of iron and the temperature of the Earth's core. *Geophys. J. R. Astr. Soc.* **85**, 315-28.

(1987). On Poisson's ratio and the composition of the Earth's lower mantle. *Phys. Earth Planet. Interiors* **46**, 357-68.

(1988a). Transport properties of liquid metals and viscosity of the Earth's core. *Geophys. J.* **92**, 99-105.

(1988b). The rheological approach to the viscosity of planetary mantles: A critical assessment. In *The physics of planets,* S. K. Runcorn, ed. New York: Wiley, 161-71.

(1989). Lindemann law and the melting temperature of perovskites. *Phys. Earth Planet. Interiors* **54**, 364-9.

Poirier, J. P., and Liebermann, R. C. (1984). On the activation volume for creep and its variation with depth in the Earth's lower mantle. *Phys. Earth Planet. Interiors* **35**, 283-93.

Poirier, J. P., and Madon, M. (1979). Transmission electron microscopy of natural $(Mg_{0.74}Fe_{0.26})_2SiO_4$ spinel. *EOS Trans. Am. Geophys. Union* **60**, 370.

Poirier, J. P.; Peyronneau, J.; Gesland, J. Y.; and Brébec, G. (1983). Viscosity and conductivity of the lower mantle; an experimental study on a $MgSiO_3$ perovskite analogue, $KZnF_3$. *Phys. Earth Planet. Interiors* **32**, 273-87.

Poirier, J. P.; Peyronneau, J.; Madon, M.; Guyot, F.; and Revcolevschi, A. (1986). Eutectoïd phase transformation of olivine and spinel into perovskite and rock salt structures. *Nature* **321**, 603-5.

Price, G. D.; Putnis, A.; Agrell, S. O.; and Smith, D. G. W. (1983). Wadsleyite, natural β-$(MgFe)_2SiO_4$ from the Peace River meteorite. *Can. Mineralogist* **21**, 29-35.

Prigogine, I. (1962). *Introduction to thermodynamics of irreversible processes.* New York: Wiley.

Putnis, A., and McConnell, J. D. C. (1980). *Principles of mineral behaviour.* New York: Elsevier.

Putnis, A., and Price, G. D. (1979). High-pressure Mg_2SiO_4 phases in the Tenham chondrite meteorite. *Nature* **280**, 217-18.

Quareni, F., and Mulargia, F. (1988). The validity of the common approximate expressions for the Grüneisen parameter. *Geophys. J.* **93**, 505-19.

Ramakrishnan, J.; Boehler, R.; Higgins, G.; and Kennedy, G. (1978). Behavior of Grüneisen's parameter of some metals at high pressure. *J. Geophys. Res.* **83**, 3535-8.

Ramsey, W. H. (1949). On the nature of the Earth's core. *Mon. Not. R. Astr. Soc. Geophys. Suppl.* **5**, 409-26.

Rao, C. N. R., and Rao, K. J. (1978). *Phase transitions in solids*. New York: McGraw-Hill.

Reddy, K. P. R.; Oh, S. M.; Major, L. D., Jr.; and Cooper, A. R. (1980). Oxygen diffusion in forsterite. *J. Geophys. Res.* **85**, 322–6.

Richet, P. (1988). Superheating, melting and vitrification through decompression of high pressure minerals. *Nature* **331**, 56–8.

Richet, P., and Bottinga, Y. (1986). Thermochemical properties of silicate glasses and liquids. *Rev. Geophys.* **24**, 1–25.

Richet, P.; Mao, H. K.; and Bell, P. M. (1988). Static compression and equation of state of CaO to 1.35 Mbar. *J. Geophys. Res.* **93**, 15279–88.

Rigden, S. M.; Jackson, I.; Niesler, H.; Ringwood, A. E.; and Liebermann, R. C. (1988). Pressure dependence of the elastic wave velocities for Mg_2GeO_4 spinel. *Geophys. Res. Lett.* **15**, 605–8.

Ringwood, A. E. (1967). The pyroxene–garnet transformation in the Earth's mantle. *Earth Planet. Sci. Lett.* **2**, 255–63.

(1970). Phase transformations and the constitution of the mantle. *Phys. Earth Planet. Interiors* **3**, 109–55.

(1975). *Composition and petrology of the Earth's mantle*. New York: McGraw-Hill.

(1977). Composition of the core and implications for origin of the Earth. *Geochemical J.* **11**, 111–35.

(1979). Composition and origin of the Earth. In *The Earth, its origin, structure and evolution*, M. W. McElhinny, ed. New York: Academic, 1–58.

(1989). Significance of the Mg/Si terrestrial ratio. *Earth Planet Sci. Lett.* **95**, 1–7.

Ringwood, A. E., and Irifune, T. (1988). Nature of the 650-km seismic discontinuity: implications for mantle dynamics and differentiation. *Nature* **331**, 131–6.

Ringwood, A. E., and Major, A. (1966). Synthesis of Mg_2SiO_4–Fe_2SiO_4 spinel solid solutions. *Earth Planet Sci. Lett.* **1**, 241–5.

(1970). The system Mg_2SiO_4–Fe_2SiO_4 at high pressures and temperatures. *Phys. Earth Planet. Interiors* **3**, 89–108.

(1971). Synthesis of majorite and other high pressure garnets and perovskites. *Earth Planet. Sci. Lett.* **12**, 411–18.

Rivier, N., and Duffy, D. M. (1982). On the topological entropy of atomic liquids and the latent heat of fusion. *J. Phys. C* **15**, 2867–74.

Robie, R. A., and Edwards, J. L. (1966). Some Debye temperatures from single-crystal elastic constant data. *J. Appl. Phys.* **37**, 2569–663.

Ross, M. (1969). Generalized Lindemann melting law. *Phys. Rev.* **184**, 233–42.

Ross, N. L., and Hazen, R. M. (1989). Single crystal X-ray diffraction study of $MgSiO_3$ perovskite from 77 to 400 K. *Phys. Chem. Minerals* **16**, 415–20.

Roufosse, M. C., and Jeanloz, R. (1983). Thermal conductivity of minerals at high pressure: The effect of phase transitions. *J. Geophys. Res.* **88**, 7399–409.

Roufosse, M. C., and Klemens, P. G. (1974). Lattice thermal conductivity of minerals at high temperatures. *J. Geophys. Res.* **79**, 703–5.

Ruoff, A. L. (1965). Mass transfer problems in ionic crystals with charge neutrality. *J. Appl. Phys.* **36**, 2903–7.

(1967). Linear shock-velocity–particle-velocity relationship. *J. Appl. Phys.* **38**, 4976–80.

Ryerson, F. J. (1987). Diffusion measurements: experimental methods. In *Methods of experimental physics, geophysics,* C. G. Sammis and T. L. Henyey, eds. New York: Academic, **24 A**, 89–130.

Salter, L. (1954). The Simon melting equation. *Philos. Mag.* **54**, 369–78.

Samara, G. A. (1967). Insulator to metal transition at high pressure. *J. Geophys. Res.* **72**, 671–8.

Sammis, C. G.; Smith, J. C.; and Schubert, G. (1981). A critical assessment of estimation methods for activation volume. *J. Geophys. Res.* **86**, 10707–18.

Sawamoto, H. (1987). Phase diagram of $MgSiO_3$ at pressures up to 24 GPa and temperatures up to 2200°C: Phase stability and properties of tetragonal garnet. In *High-pressure research in mineral physics,* M. H. Manghnani and Y. Syono, eds. Washington, D.C.: American Geophysical Union, 209–19.

Sawamoto, H.; Weidner, D. J.; Sasaki, S.; and Kumazawa, M. (1984). Single crystal elastic properties of modified spinel (Beta) phase of magnesium orthosilicate. *Science* **224**, 749–51.

Saxton, H. J., and Sherby, O. D. (1962). Viscosity and atomic mobility in liquid metals. *Trans. ASM* **55**, 826–43.

Schock, R. N.; Duba, A. G.; and Shankland, T. J. (1989). Electrical conduction in olivine. *J. Geophys. Res.* **94**, 5829–39.

Schreiber, E. and Anderson, O. L. (1970). Properties and composition of lunar materials: Earth analogies. *Science* **168**, 1579–80.

Secco, R. A., and Schloessin, H. H. (1989). The electrical resistivity of solid and liquid Fe at pressures up to 7 GPa. *J. Geophys. Res.* **94**, 5887–94.

Shankland, T. J. (1972). Velocity–density systematics: Derivation from Debye theory and the effect of ionic size. *J. Geophys. Res.* **77**, 3750–8.

(1977). Elastic properties, chemical composition and crystal structure of minerals. *Geophys. Surveys* **3**, 69–100.

Shankland, T. J., and Brown, J. M. (1985). Homogeneity and temperatures in the lower mantle. *Phys. Earth Planet. Interiors* **38**, 51–8.

Shankland, T. J.; Nitsan, U.; and Duba, A. G. (1979). Optical absorption and radiative heat transport in olivine at high temperature. *J. Geophys. Res.* **84**, 1603–10.

Shapiro, J. N., and Knopoff, L. (1969). Reduction of shock-wave equations of state to isothermal equations of state. *J. Geophys. Res.* **74**, 1435–8.

Sherman, D. M. (1989). The nature of the pressure-induced metallization of FeO and its implication to the core–mantle boundary. *Geophys. Res. Lett.* **16**, 515–18.

Shimoji, M., and Itami, T. (1986). *Atomic transport in liquid metals.* Switzerland: TransTech Publications.

Shockley, W. (1952). Dislocation models of grain boundaries. *L'Etat Solide.* Brussels: 9ème Conseil de Physique Solvay.

Simmons, G., and Wang, H. (1971). Single crystal elastic constants and calculated aggregate properties: a Handbook. Cambridge, MA: MIT Press.

Simon, F., and Glatzel, G. (1929). Bemerkungen zur Schmelzdruckkurve. *Z. Anorg. Allg. Chem.* **178**, 309–16.

Slater, J. C. (1939). *Introduction to chemical physics.* New York: McGraw-Hill.

Sneeringer, M.; Hart, S. R.; and Shimizu, N. (1984). Strontium and samarium diffusion in diopside. *Geochim. Cosmochim. Acta* **48**, 1589–608.

Spiliopoulos, S., and Stacey, F. D. (1984). The Earth's thermal profile: Is there a mid-mantle thermal boundary layer? *J. Geodynamics* **1**, 61–77.

Stacey, F. D. (1972). Physical properties of the Earth's core. *Geophys. Surveys* **1**, 99–119.

(1977a). A thermal model of the Earth. *Phys. Earth Planet. Interiors* **15**, 341–8.

(1977b). Applications of thermodynamics to fundamental Earth physics. *Geophys. Surveys* **3**, 175–204.

Stacey, F. D., and Irvine, R. D. (1977a). Theory of melting: Thermodynamic basis of Lindemann law. *Aust. J. Phys.* **30**, 631–40.

(1977b). A simple dislocation theory of melting. *Aust. J. Phys.* **30**, 641–6.

Stacey, F. D.; Brennan, B. J.; and Irvine, R. D. (1981). Finite strain theories and comparison with seismological data. *Geophys. Surveys* **4**, 189–232.

Stebbins, J. F.; Carmichael, I. S. E.; and Moret, L. K. (1984). Heat capacities and entropies of silicate liquids and glasses. *Contrib. Min. Petrol.* **86**, 131–48.

Steinberg, D. J. (1981). The temperature independence of Grüneisen's gamma at high temperature. *J. Appl. Phys.* **52**, 6415–17.

Stern, E. A., and Zhang, K. (1988). Local premelting about impurities. *Phys. Rev. Lett.* **60**, 1872–5.

Stevenson, D. J. (1981). Models of the Earth's core. *Science* **214**, 611–19.

Stishov, S. M.; Makarenko, I. N.; Ivanov, V. A.; and Nikolaenko, A. M. (1973). On the entropy of melting. *Physics Lett.* **45A**, 18.

Strong, H. M.; Tuft, R. E.; and Hanneman, R. E. (1973). The iron fusion curve and the γ–δ–l triple point. *Metallurgical Trans.* **4**, 2657–61.

Stumpe, R.; Wagner, D.; and Bäuerle, D. (1983). Influence of bulk and interface properties on the electric transport in ABO_3 perovskites. *Phys. Stat. Sol.* **75**, 143–54.

Suito, K. (1977). Phase relations of pure Mg_2SiO_4 up to 200 kilobars. In *High-pressure research,* M. H. Manghnani and S. Akimoto, eds. New York: Academic, 255–66.

Sumino, Y., and Anderson, O. L. (1984). Elastic constants of minerals. In *Handbook of physical properties of rocks.* Boca Raton: CRC Press, 39–138.

Sumino, Y.; Anderson, O. L.; and Suzuki, I. (1983). Temperature coefficients of elastic constants of single crystal MgO between 80 and 1300 K. *Phys. Chem. Minerals* **9**, 38–47.

Sutherland, W. (1890). A new periodic property of the elements. *Philos. Mag.* **30**, 318–23.

(1891). A kinetic theory of solids, with an experimental introduction. *Philos. Mag.* **32**, 31–43.

Suzuki, H. (1983). Dislocation models of liquids. Basic concepts and phenomena related to atomic motions. In *Topological disorder in condensed matter,* F. Yonegawa and T. Ninomiya, eds. Berlin: Springer.

Suzuki, I.; Ohtani, E.; and Kumazawa, M. (1979). Thermal expansion of γMg_2SiO_4. *J. Phys. Earth* **27**, 53–61.

Suzuki, T.; Akimoto, S.; and Fukai, Y. (1984). The system iron–enstatite–water at high pressures and temperatures – formation of iron hydride and some geophysical implications. *Phys. Earth Planet. Interiors* **36**, 135–44.

Swalin, R. A. (1959). On the theory of self-diffusion in liquid metals. *Acta Metall.* **7**, 736–40.

Takahashi, E. (1986). Melting of dry peridotite KLB-1 up to 14 GPa: Implications on the origin of the peridotitic upper mantle. *J. Geophys. Res.* **91**, 9367–82.

Takeuchi, H., and Kanamori, H. (1966). Equations of state of matter from shock wave experiments. *J. Geophys. Res.* **71**, 3985–94.

Tallon, J. L. (1980). The entropy change on melting of simple substances. *Physics Lett.* **76A**, 139–42.

Tallon, J. L., and Robinson, W. H. (1977). A mechanical instability hypothesis for melting in the alkali halides. *Philos. Mag.* **36**, 741–51.

Tallon, J. L.; Robinson, W. H.; and Smedley, S. I. (1977). A melting criterion based on the dilatation dependence of shear moduli. *Nature* **266**, 337–8.

Tannhauser, D. S. (1962). Conductivity in iron oxides. *J. Phys. Chem. Solids* **23**, 25–34.

Tritton, D. J. (1977). *Physical fluid dynamics.* New York: Reinhold.

Tuller, H. L., and Nowick, A. S. (1977). Small polaron electron transport in reduced CeO_2 single crystals. *J. Phys. Chem. Solids* **38**, 859–67.

Turcotte, D. L., and Schubert, G. (1982). *Geodynamics.* New York: Wiley.

Ubbelohde, A. R. (1978). *The molten state of matter.* London: Wiley.

Urakawa, S.; Kato, M.; and Kumazawa, M. (1987). Experimental study on the phase relations in the system Fe-Ni-O-S up to 15 GPa. In *High-pressure research in mineral physics,* M. H. Manghnani and Y. Syono, eds. Washington, D.C.: American Geophysical Union, 95–111.

Vaidya, S. N., and Gopal, E. S. R. (1966). Melting law at high pressures. *Phys. Rev. Lett.* **17**, 635–6.

Van Liempt, J. (1935). Die Berechnung der Auflockerunswärme der Metalle aus Rekristallisationsdaten. *Z. Phys.* **96**, 534–41.

Vashchenko, V. Y., and Zubarev, V. N., (1963). Concerning the Grüneisen constant. *Soviet Phys. Solid State* **5**, 653–5.

Vaughan, P. J.; Green, H. W.; and Coe, R. S. (1982). Is the olivine–spinel phase transformation martensitic? *Nature* **298**, 357–8.

Verhoogen, J. (1980). *Energetics of the Earth.* Washington, D.C.: National Academy of Sciences.

Von Horst Bester, and Lange, K. W. (1972). Abschätzung mittlerer Werte für die Diffusion von Kohlenstoff, Sauerstoff, Wasserstoff, Stickstoff und Schwefel in festem und flüssigen Eisen. *Archiv für das Eisenhüttenweisen* **3**, 207–13.

Wall, A., and Price, G. D. (1989). Electrical conductivity of the lower mantle, a molecular dynamics simulation of $MgSiO_3$. *Phys. Earth Planet. Interiors* **58**, 192–204.

Walzer, U. (1982). Volume dependence of melting temperature at high pressure and its relation to a new dimensionless quantity. *Exper. Technik der Physik* **31**, 33–51.

Wang, C., Y. (1968a). Equation of state of periclase and Birch's relationship between velocity and density. *Nature* **218**, 74–6.

(1968b). Constitution of the lower mantle as evidenced from shock-wave data for some rocks. *J. Geophys. Res.* **73**, 6459–76.

(1969). Equation of state of periclase and some geophysical implications. *J. Geophys. Res.* **74**, 1451–7.

(1970). Density and constitution of the mantle. *J. Geophys. Res.* **75**, 3264–84.

(1972a). A simple Earth model. *J. Geophys. Res.* **77**, 4318–29.

(1972b). Temperatures in the lower mantle. *Geophys J. R. Astr. Soc.* **27**, 29–36.

(1978). Elastic constants systematics. *Phys. Chem. Minerals* **3**, 251–61.

Watt, J. P. (1988). Elastic properties of polycrystalline minerals: comparison of theory and experiment. *Phys. Chem. Minerals* **15**, 579–87.

Watt, J. P.; Davies, G. F.; and O'Connell, R. J. (1976). The elastic properties of composite materials. *Rev. Geophys. Space Phys.* **14**, 541–63.

Weertman, J. (1970). The creep strength of the Earth's mantle. *Rev. Geophys. Space Phys.* **8**, 145–68.

(1978). Creep laws for the mantle of the Earth. *Phil. Trans. R. Soc. London* **A228**, 9–26.

Weidner, D. J. (1986). Mantle models based on measured physical properties of minerals. In *Chemistry and physics of terrestrial planets,* S. K. Saxena, ed. Berlin: Springer, 251–74.

(1987). Elastic properties of rocks and minerals. In *Methods of experimental physics, geophysics,* C. G. Sammis and T. L. Henyey, eds. New York: Academic, **24 A**, 1–30.

Weidner, D. J., and Ito, E. (1985). Elasticity of $MgSiO_3$ in the ilmenite phase. *Phys. Earth Planet. Interiors* **40**, 65–70.

Weidner, D. J.; Sawamoto, H.; Sasaki, S.; and Kumazawa, M. (1984). Single crystal elastic properties of the spinel phase of Mg_2SiO_4. *J. Geophys. Res.* **89**, 7852–60.

Wells, A. F. (1985). *Structural inorganic chemistry.* Oxford: Oxford University Press.

White, G. K., and Anderson, O. L. (1966). Grüneisen parameter of magnesium oxide. *J. Appl. Phys.* **37**, 430–2.

Will, G.; Cemic, L.; Hinze, E.; Seifert, K. F.; and Voigt, R. (1979). Electrical conductivity measurements on olivines and pyroxenes under defined thermodynamic activities as a function of temperature and pressure. *Phys. Chem. Minerals* **4**, 189–97.

Will, G.; Hoffbauer, W.; Hinze, E.; and Lauterjung, J. (1986). The compressibility of forsterite up to 300 kbar measured with synchrotron radiation. *Physica* **139–40B**, 193–7.

Williams, Q.; Jeanloz, R.; Bass, J.; Svendsen, B.; and Ahrens, T. J. (1987). The melting curve of iron to 250 gigapascals: A constraint on the temperature at the Earth's center. *Science* **236**, 181–2.

Williams, Q.; Knittle, E.; and Jeanloz, R. (1987). High temperature experiments on liquid iron alloys: Applications to the Earth's core. *EOS trans. AGU* **68**, 1493.

Williamson, E. D., and Adams, L. H. (1923). Density distribution in the Earth. *J. Washington Acad. Sci.* **13**, 413–28.

Wolf, G. H., and Bukowinski, M. S. T. (1985). Ab initio structural and thermoelastic properties of orthorhombic $MgSiO_3$. *Geophys. Res. Lett.* **12**, 809–12.

(1987). Theoretical study of the structural properties and equations of state of $MgSiO_3$ and $CaSiO_3$ perovskites: implications for lower mantle composition. In *High-pressure research in mineral physics,* M. H. Manghnani and Y. Syono, eds. Washington, D.C.: American Geophysical Union, 313–31.

Wolf, G. H., and Jeanloz, R. (1984). Lindemann melting law: Anharmonic correction and test of its validity for minerals. *J. Geophys. Res.* **89**, 7821–35.

Xu, J. A.; Mao, H. K.; and Bell, P. M. (1986). High-pressure ruby and diamond fluorescence: Observations at 0.21 to 0.55 terapascal. *Science* **232**, 1404–6.

Yagi, T., and Akimoto, S. (1974). Electrical conductivity jump produced by the α–β–δ transformations in Mn_2GeO_4. *Phys. Earth Planet. Interiors* **8**, 235–40.

Yagi, T.; Akaogi, M.; Shimomura, O.; Suzuki, T.; and Akimoto, S. (1987). In situ observation of the olivine–spinel transformation in Fe_2SiO_4 using synchrotron radiation. *J. Geophys. Res.* **92**, 6207–13.

Yagi, T.; Akaogi, M.; Shimomura, O.; Tamai, H.; and Akimoto, S. (1987). High pressure and high temperature equations of state of majorite. In *High-pressure research in mineral physics,* M. H. Manghnani and Y. Syono, eds. Washington, D.C.: American Geophysical Union, 141–7.

Yagi, T.; Mao, H. K.; and Bell, P. M. (1982). Hydrostatic compression of perovskite-type $MgSiO_3$. In *Advances in physical geochemistry,* S. K. Saxena, ed. New York: Springer, 317–25.

Yamashita, J., and Kurosawa, T. (1958). On electronic current in NiO. *J. Phys. Chem. Solids* **5**, 34–43.

Yeganeh-Haeri, A.; Weidner, D. J.; and Ito, E. (1989). Elasticity of $MgSiO_3$ in the perovskite structure. *Science* **243**, 787–9.

Young, C. J., and Lay, T. (1987). The core–mantle boundary. *Ann. Rev. Earth Planet. Sci.* **15**, 25–46.

Zharkov, V. N., and Kalinin, V. A. (1971). *Equations of state for solids at high pressures and temperatures.* New York: Consultants Bureau.

Zharkov, V. N.; Karpov, P. B.; and Leontjeff, V. V. (1985). On the thermal regime of the boundary layer at the bottom of the mantle. *Phys. Earth Planet. Interiors* **41**, 138–42.

Ziman, J. M. (1965). *Principles of the theory of solids.* Cambridge: Cambridge University Press.

Zwanzig, R. (1983). On the relation between self-diffusion and viscosity of liquids. *J. Chem. Phys.* **79**, 4507–8.

Zwikker, C. (1954). *Physical properties of solid materials.* Oxford: Pergamon Press.

Index

Pages where the topic is explained in depth are indicated in **boldface** type.